THE HARD CORPS

21ST CENTURY
LEADERSHIP DEVELOPMENT

THE HARD CORPS

21ST CENTURY
LEADERSHIP DEVELOPMENT

STEWART W. HUSTED

MARINER
PUBLISHING
Buena Vista, VA

1 3 5 7 9 10 8 6 4 2

Library of Congress Control Number: 2009925478

The Hard Corps
21st Century Leadership Development
Stewart W. Husted
Includes Index and Bibliographical References

p. cm.
1. Leadership 2. Leadership Development 3. Military Training
4. Virginia Military Institute
5. Military Colleges 6. Executive Ability 7. Experiential Learning

I. Husted, Stewart Winthrop, 1946— II. Title.
ISBN 13: 978-0-9820172-8-9 (softcover : alk. Paper)
ISBN 10: 0-9820172-8-6

Book design by Tracy Lee Staton
Cover design by Melanie M. Wills

Mariner Publishing
A division of
Mariner Media, Inc.
131 West 21st ST.
Buena Vista, VA 24416
Tel: 540-264-0021
http://www.marinermedia.com

Printed in the United States of America

This book is printed on acid-free paper meeting the requirements of the American Standard for Permanence of Paper for Printed Library Materials.

The Compass Rose and Pen are trademarks of Mariner Media, Inc.

DEDICATION

To all cadets, alumni, faculty and staff who have served in the Global War on Terror and a special tribute to those VMI alumni who gave their lives during this war for our nation ... true citizen-soldiers.

Captain James C. Edge 1996, USMC
Captain Lowell T. Miller II 1993, ANG
Major Paul R. Syverson III 1993, USA Special Forces
Sergeant Ryan E. Doltz 2000, ANG
Captain John R. Teal 1994, USA
1st Lt. Joshua C. Hurley 2001, USA
Captain Luke C. Wullenwaber 2002, USA
Lt. Commander David L. Williams 1992, USN
First Sergeant Luke J. Mercardante, honorary BR 2007, USMC
Specialist William L. McMillan III, 2008, USA
Charles W. Mathers 1962, civilian in World Trade Tower One

And to my son, Captain Ryan W. Husted (USA), who served 15 months in Iraq as a Black Hawk pilot and commander of Alpha Company (Air Assault), 1st Infantry Division, Combat Aviation Brigade.

TABLE OF CONTENTS

AUTHOR'S NOTE

Turning onto Lexington's Letcher Avenue in August of 2002, I saw the beginning and end of my life's journey. Over 50 years earlier, the route began on Letcher Avenue in a small basement apartment on Washington & Lee University's faculty row. It seemed so long ago, almost as if never happened. My drive up the avenue took me past the Lee Chapel, faculty row, and through the beautiful campus of Washington and Lee University. Straight ahead were the entrance gates to my destination-- the campus of the Virginia Military Institute. I was traveling to VMI to join the faculty and become a member of what is known as the VMI family.

Driving slowly at the posted 15 mph, my mind flashed back to the earlier years when my mother walked me up the same block to the VMI Parade Field. To pass time while waiting for my father, a chemistry professor at W&L, we watched the weekly parades, which included VMI's last cavalry unit. Along the border of the parade field, a long row of old brown, weathered benches, where we sat, still rest--each one dedicated as a memorial to someone associated with the Institute. To me they represent a special VMI memory from my past.

The Hard Corps is part-memoir with a dash of VMI history included; but first and foremost it is a leadership book based on my reflections and leadership experiences over the past 40 years as they apply to VMI. Those experiences include seven years as a VMI faculty member; my earlier years as a cadet at Virginia Tech; service as an Army officer (USAR retired); tenure as a business school faculty member and dean and service as the founder and president of a non-profit leadership training organization.

To some, it may seem odd that a non-VMI alumnus would write a book about VMI's role in developing leaders who as citizen-soldiers serve Virginia, the nation, and even the world.

Sometimes the most insightful books are written by "outsiders," who provide a fresh look at how things are done. I served and observed VMI from 2002-2009 as a full-time member of the Economics and Business faculty. Before retiring, I witnessed some of the best and worst of cadet behavior and leadership; however, this book is not a kiss-and-tell memoir for there would be no point. Every college or university has it warts and inside gossip. VMI is no different, but for the most part the Institute runs as efficiently and effectively as any institution I've ever served ... actually much more so. It is a college that all of America should be proud to support. Its mission and goals are All-American.

In the forthcoming chapters, I will explain why our nation needs military colleges like VMI; focus on the leader development program used by VMI to train and educate citizen-soldiers; and illustrate how this system applies to developing civilian as well as military leaders. The book will attempt to directly or indirectly answer the questions: How does VMI produce so many successful graduates? How does VMI develop leaders of integrity? What can others aspiring to leadership positions learn from the VMI leader development program?

Some readers may not recognize the VMI described in these pages. That is understandable, but alumni need to realize that despite VMI's deep traditions, nothing stays or lasts forever, even at the Institute, as many call it. Administrators estimate that it takes only five to seven years to create a near complete change in customs ranging from uniforms worn to new Breakout (induction) practices. Thus, VMI continues to evolve. It is not the same place it was before or after the Civil War, after World War II or the Korean War. For instance, cadets returning today from the Global War on Terrorism can't imagine that VMI ever allowed students returning from World War II and the Korean War to live off Post or, heaven forbid, to be married. Neither is it the same place it was before minorities were admitted in 1968 or women in 1996. Certainly one could easily argue that through change, VMI is a stronger institution. It is an institution that is constantly striving for excellence, which is one reason it is continually ranked in the top three public liberal arts colleges in America.

While each of these changes makes the VMI fabric a little stronger, like fabric from different bolts, each class is always a little different shade ...in this case a slightly different shade of gray. Thus, the "VMI experience" will always be different things to different classes. At its core, though, will always be a common series of experiences, but uniquely different in some way for each new Rat class (freshman) over a four-year period. The experience is tailored to their class of Brother Rats by their dykes (mentors), faculty and staff members, and others who shape it.

Finally, this book will identify the most important leadership traits, actions, and experiences needed to become a successful 21st century leader. It will profile not only cadets, but also faculty, administration, and alumni --- all whose successes and leadership impacted their state, nation and world. Lastly, the reader will examine the source, process, and practice of leader development at VMI.

Stewart W. Husted, Ph.D.
Virginia Military Institute
February 4, 2009

ACKNOWLEDGEMENTS

Many people are responsible for making this book a reality. Because no book dealing with so many historical facts and stories about VMI had been written in over 20 years, it became apparent that a whole generation of alumni, faculty and staff history needed to be documented and many people interviewed to illustrate the current VMI leadership development system. Therefore, I am grateful for the assistance given to me by the following individuals and organizations:

The VMI Corps of Cadets and the over 175 cadets who participated in my business leadership course from 2002-2008; Brigadier General (ret.) Casey Brower whose insights helped create the structure of the book; Colonel Keith Gibson, who checked the manuscript for historical errors; Colonel Vernon Beitzel, who provided enrollment data and reviewed the manuscript; Colonel Karen Gutermuth, who insured accuracy for information about the honor system; Donnie White, who reviewed the manuscript and provided additional sports stories; Thomas Trumps, who lent crucial support and information for the project; Donna Potter, who provided many secretarial skills; the excellent copy editing and suggestions of Nina Salmon of the VMI Writing Center and, and the diligent editing and support of my publisher, Andrew Wolfe and his staff at Mariner Media.

Others who were interviewed or provided assistance included: Adam Volant, Colonel Tom Davis, Colonel Floyd Duncan, Colonel James Tubbs, Colonel Gordon Caulkins and his cadet and faculty staff of Rat Challenge, Colonel Cliff West, Dr. Larry Bland, 1st Lieutenant Sal Sferrazza, Captain Matt Thompson, Captain Ryan Husted, William Diel, Jr., Gerald Quirk, Travis Russell, Scott Sayre, Donna Potter, Cadets B. Finney Kimsey, Jake Jackson, Jonathon Faff, Eric Hunter, Paul Childrey, Alex Snyder, and the helpful staff at the Preston Library and Marshall Foundation.

INTRODUCTION
IN SEARCH OF TOMORROW'S LEADERSHIP

"We are facing military challenges today unlike in the past. Our nation needs to rely on young leaders such as you [VMI] new young leaders to do what you have done in the past --- to defend our nation here and abroad, to conduct yourself honorably with or without specific guidelines, and to enable our nation to survive in some of the most challenging times we have ever faced."

Sandra Day O'Connor,
Supreme Court Justice -
Address to the VMI Corps of Cadets
March 26, 2008

This is a book about leadership and how one institution of higher education develops leaders through a series of educational and co-curricular experiences. It is also the story of leaders from all walks of life, who are or were part of the Virginia Military Institute (VMI) family of cadets, alumni, faculty, and staff. These stories are about individuals whose leadership made a positive difference in the lives of others. To look at their physical appearance may not provide you with clues to their roles as leaders: some wear suits, some clerical collars, some military uniforms of all ranks, and some the scrubs of physicians or the lab coats of scientists. While we may not be able to recognize them by their appearance, it is easy to recognize these leaders by their actions. Supreme Court Justice Potter Stewart, when asked to define obscenity, once responded that he couldn't define it, but he knew it when he saw it. So it is with leadership...hard to define, but easier to observe.

Leadership: An Art or Science?

Some contend that leaders are born, not made. They believe leadership is all about charisma and personality. Some of these individuals would also have us believe that leadership is nothing more than good common sense … just an application of good business practices to solve the challenges of an organization. In other words, leadership is learned through trial and error. Conversely, others maintain that leadership is a science, growing out of the field of psychology and developed as a subject that can be studied and taught to potential leaders. By studying leadership and its theories and practices, we can avoid the mistakes of other leaders. In reality, leadership is both an art and a science. It is a very complex subject, which requires research in order to understand fully the nuances of human behavior under changing and challenging situations. By understanding both the theories of leadership and examples of positive leadership, students can begin to see how their behavior as leaders affects those they seek to influence.

Is America Experiencing a Leadership Crisis?

There are more definitions of leadership than anyone wants to count, but one fact remains true to all. Leadership is a form of social influence. By influencing the behavior of others, leaders initiate and guide followers through a process of effective change to accomplish their common goals. Leaders often leave a timeless mark on people, organizations, nations and even civilizations.[1] Leadership is also a process of discovering new ideas and one's own values, qualities and weaknesses, and visions. This is reinforced by Ann Fudge, CEO of Young & Rubicam Brand, who tells us, "All of us have a spark of leadership within us, whether it is in business, in government, or as a non-profit volunteer. The challenge is to understand ourselves well enough to serve others."[2] Thus, the study of leadership is also a process of self-discovery.

In 2001, many young people, from Pat Tilman, NFL star, to Sal Sferrazza, a freshman photography major at SUNY Farmington, started a process of self-discovery and reexamination of their values after the 9/11 attacks. Young adults and college students who rallied to the call of patriotism were not your typical military recruit, but individuals who realized that the America they knew would rapidly change. Consequently, they began to ask themselves key life questions. How would I live my life if I had a short time to live? How can I use my God-given talents to serve others? How can I make a difference in these chaotic times? How do I fit into the greater global picture?

Now more than ever we need experienced leaders of character who will do the right thing in these challenging and difficult times. Our nation does not need leaders who choose popular courses of action demanded by the press and others. True leadership is not about public polls. The constant change of positions on issues by many political leaders is likely a by-product of polling trends and may be part of our problem in selecting our best, brightest, and most experienced leaders.

According to David Gergen of Harvard University's Center for Public Leadership at the John F. Kennedy School of Government, America is in a leadership crisis.[3] The American public is reeling from a series of poor leaders and leadership decisions. The public's confidence and trust in our leaders, corporate, government, and military is quickly eroding. In each of 11 fields surveyed by a *U.S. News* & *World Report* and Harvard University survey, "no more than 40 percent of Americans said they had a great deal of confidence in their leaders," while only 39 percent said leaders had high ethical standards.[4]

In the past few years, we witnessed corporate CEOs, who engaged in unethical business practices ranging from inside trading to fraud; in government, we witnessed bad civilian leadership decisions, when Congressmen accepted bribes from foreign officials, a governor solicited call girls, and presidents led us to wars, which failed to achieve victory, despite having the best trained and equipped military in the world; and in the military, we observed generals and others who attempted to cover up the horrors of friendly fire and prison detainee abuse. The impact of failed leadership in all sectors of society is felt everywhere. Pensions evaporate, lives are lost, families destroyed, jobs are gone forever as companies cease to exist. These past and present leaders failed the only valid test of leadership: achieving sustainable results over time. In doing so, they also failed their followers and created an environment of distrust.

Searching for Role Models

A new definition of leadership is emerging for the 21st century. Proponents include military leaders as well as CEOs. According to Harvard professor and former Medtronic CEO Bill George, "The military-manufacturing model of leadership that worked so well 50 years ago doesn't get the most out of people today. People are too well informed to adhere to a set of rules or to simply follow a leader over a distant hill. They want to be inspired to a greater purpose."[5] This statement has tremendous implications for VMI, as it is an institution about rules and regulations. VMI must think about the roles U.S. soldiers are performing in the Global War on Terrorism.

Diane Sawyer and Charles Gibson broadcast
Good Morning America live from VMI
on Veterans Day, November 11, 2002.

Retired Supreme Court Justice O'Connor told cadets at VMI, when receiving the Harry E. Byrd Jr. 1935 Public Service Award in 2008 (March 26), "The nation has placed an enormous burden upon our Armed Forces. We have asked them to be our soldiers and our statesmen, to be our combatants and our conscience. This burden has been placed upon them with only limited guidance. But the history and honor of this institution suggests that if anyone can bear this responsibility, it is the cadets of the Virginia Military Institute, and your previous graduates."[6]

Today decisions are being made at the platoon level and lower that impact villages and tribal areas. Our young military leaders are told to sit down with the enemy and talk, to create alliances and partnerships to win the peace, and to rebuild communities. Hardly, the kick down the door and shoot first approach initially used in Iraq. Leaders must be willing and able to adapt and serve a higher purpose of service. The leaders of this century must be able to instill a call to service and sacrifice in their followers. The leader's character becomes a personal GPS system ... a constant navigational system to know where they are in life and what direction to take during challenging times. Our character guides us to do the right thing, which becomes harder when there are no sets of directions or regulations dictating prescribed actions.

As those of today's Millennium Generation seek their role models, they won't find a George C. Marshall or a Martin Luther King Jr. Instead, their search could likely lead to wannabe leaders, who are quickly destroyed by the media. If we believe the mass media, there are flaws woven into the fabric of every leader. The media often rushes to uncover and judge the moral or intellectual fiber of many potential leaders; however, Americans can be forgiving of these flaws if a leader can inspire us and make a difference through positive change and social impact. For example, former New York City Mayor Rudy Giuliani took a beating in the press for leaving his wife and children for another woman. His political future and ability to lead as mayor were quickly coming to a close. From the ashes of New York City and 9/11, Giuliani rose to lead the city and to become a presidential candidate in 2008. He demonstrated the true grit of strong leadership and management skills coupled with compassion for others. Former President Jimmy Carter is not noted by historians as being a very good president; however, once out of office his agenda of public service to causes of human justice, peace and social good won him the Nobel Peace Prize.

Statue of General George C. Marshall, class of 1902.
Marshall was the Chief of Staff of the Army during World War II
and recipient of the Nobel Peace Prize.

Today's young people have their own set of leader and hero role models, and perhaps a different set of standards for identifying those leaders who inspire them. According to a recent U.S. Chamber of Commerce survey, these role models are very important; they rank second (behind experience) on how people learn to lead.[7] However, I sincerely hope that their standards for role models include good character. As an educator, I believe we should ensure that our students recognize the value and importance of good character and ethical behavior. Without these key ingredients, there can be no trust between a leader and follower. We, as the leaders and faculty of VMI, have a responsibility to provide a process or system, which is capable of developing leaders of character to serve our nation and the Commonwealth of Virginia --- a leader development system, which produces leaders who instill trust because they are known to be honorable and forthright individuals.

We must challenge every student or cadet to accept the responsibility of discovering the leadership potential within and to realize that each one can make a difference at all levels of society and organizations. Everyone can lead. Individuals, who think and speak a common code of ethical behavior, will be in high demand as our nation's cultural fabric faces more rapid change and our global environment produces an increasing number of difficult challenges to our very way of life and existence in the United States.

How Do We Recognize Authentic Leaders?

Today's authentic leaders are boldly stepping forward into the perceived leadership vacuum. The problem is that we need many more leaders as our population expands and the problems facing our nation become more complex. These new leaders such as Barack Obama represent an array of different careers and skill sets, but all are moving forward and are determined to stay the course through what could prove to be our nation's most difficult time since World War II.

Bill George and Peter Sims interviewed 125 top CEOs for their book, *True North: Discover Your Authentic Leadership.*[8] Better known examples of these executives included Warren Buffett (financial expert), Thad Allen (Commandant of the Coast Guard), Sandra Day O'Connor (retired Supreme Court justice), Michael Bloomberg (New York City mayor and education reformer), Wendy Kopp (social entrepreneur and founder of Teach for America), and A.G. Lafley (CEO of Proctor & Gamble) among other notables. These and other leaders shine as beacons in a world dimmed by cloudy ethical and moral issues. They are examples of successful leaders, who have survived challenging times in their organizations and met success through tireless effort and personal sacrifice.

Conclusion

As we examine the system of leadership development at VMI in upcoming chapters, it is important that we think of the characteristics of the real leaders identified by George and Sims in *True North.*[9] These real or "authentic" leaders:

1. Pursue their purpose with passion
2. Practice solid values
3. Lead with their hearts as well as their heads
4. Establish connected relationships
5. Demonstrate self-discipline

These are traits that I witness demonstrated every day at VMI by our cadets. I feel strongly that VMI has a tradition and a system for developing such authentic leaders, but like all models, they eventually change.

2

CORPS VALUES
THE LEADERSHIP ROLE OF THE
21ST CENTURY MILITARY COLLEGE

"One of George Washington's greatest worries was the lack of leadership academies in the United States. Your [VMI] leadership requirement goes back to the reason VMI was founded, and Washington would be pleased to see the creation of this Institute."

Michael Beschloss, author and historian,
to the Corps of Cadets-
September 18, 2007

Rick Atkinson, four-time Pulitzer Prize winning author spoke to the Corps of Cadets and "impressed upon cadets the importance of experiencing the stories of those who preceded them on the battlefield."[1] Atkinson told his listeners that he writes about war "because war is a great revealer of character. It is true of every great military leader and of every soldier in a fox hole or sailor in a gun turret."[2]

There are few better places to learn about war and character than VMI. The Post itself was a battleground during 1864, and its cadets have known war first hand from the Battle of New Market. Those 241 brave VMI cadets at New Market left a permanent impression on generations of soldiers and civilians. And so it is that like Atkinson's Liberation Trilogy, this book tries to "bring them [soldiers] to life and keep them living."[3]

Standing on High Ground

The 134-acre Virginia Military Institute stands high on a bluff overlooking the Maury River in Lexington, Virginia. Founded in 1839 as the first state-supported military college, it was also the first southern college to teach engineering and industrial chemistry. By the end of the 1850s, VMI was used as a model to create scores of other military schools and academies in the slave-holding states. From the opposite south side of Post, the historic town of Lexington can be seen a few hundred yards away. The buildings of VMI are constructed in a similar style to West Point, but instead of the gray granite that stamps the character of the Academy's gothic buildings, softer beige-colored stucco structures dominate the VMI landscape. Cadets live and attend classes in buildings registered as national historical monuments.

The Old Barracks, first occupied in 1852, stand in sharp contrast to the adjoining campus of Washington & Lee University (W&L) with its newer dorms and fraternity and sorority houses with modern suites, apartments, game rooms and lounges. The barracks, where cadets live up to five per room, were bombarded during the Civil War by Union General David Hunter's troops. It seems Hunter wanted retaliation after the VMI cadets, 241 strong, helped defeat his troops at the Battle of New Market. Only the façade, complete with patched cannon ball holes, remains; but to illustrate that VMI seeks change, a third connecting barracks housing 150 additional cadets was completed in early 2009.

The "Old" Barracks and "New Barracks"
as they stood prior to 2008.

Despite cries from many of its old grads who protested the construction of a more modern facility, the "Third" Barracks addition, is definitely not about making VMI a kinder, gentler place for cadets. There will still be no phones in rooms, no private or semi-private baths, no air-conditioning, no mini-bars with microwaves and frigs; and yes, cadets will still have to stack their racks (cots) and roll their hays (mattresses) each and every day and go down the outside stoop to use the bathroom.

The "Third Barracks" completed in 2009
includes a PX, Bookstore and Visitor's Center
at the far left side of the building.

Few visitors are permitted in the barracks area, but as an occasional visitor, I must say it reminds me of a state prison… undoubtedly the intention. After all, the site of the VMI Post was originally a military arsenal with a checkered past created by undisciplined state militiamen who guarded the site. VMI's founders (Colonel J. T. L. Preston had strong connections to W & L) proposed a military college as a way of eliminating the arsenal and returning their town to quieter times. They sought obedient students, who were subject to military order and discipline.[4] Thus, the Old Barracks' rooms open onto a central

11

courtyard with a sentry post at the center. The barracks are four levels above ground (a sub-terrain level, known as the "sink," also exists for the armory). The fourth level (stoop) is reserved for the Fourth Class, better known as Rats, while each of the other classes resides one level lower with the First Class on the ground level. At VMI everything and everybody has its place, a place reserved and preserved by 170 plus years of tradition.

Mission: Educating Leaders and Citizen Soldiers for an Uncertain Future

A college or any other worthwhile institution cannot be built alone on the foundation of its physical structures. Instead it must be built on the foundation of values and principles for which it stands. VMI, while rooted in deep values and history, strives to be "a leader and 'relevant' to the future needs of our nation."[5] Since 1839, VMI has graduated over 18,000 men and women. Of these, eleven were Rhodes Scholars and seven won the Medal of Honor; more than 265 achieved the rank of general or admiral in the military; 32 became college or university presidents, and many more became leaders and heads of Fortune 500 companies. VMI continues to graduate alumni based on a model of producing citizen-soldiers; however, it is not the purpose of VMI to provide professional training for a military career. It is the mission of VMI to educate and train cadets for useful citizenship. President Franklin Roosevelt said in 1939 when visiting VMI, "The whole history of VMI is a triumphant chronicle of the part which the citizen-soldier can play in democracy."

Senator Harry F. Byrd 1935 told the Newcomen Society in 1984 that "In honoring VMI, you really honor its product … namely, a body of men [and women] trained to lead careers of their own choosing to serve as citizen-soldiers in time of national peril …"[6]

Through the leadership of General Binford Peay III, 14th Superintendent of VMI, the Institute is striving to reexamine its vision as a military college in a world composed of traditional threats from foreign governments and new threats from terrorism.

Cadet walking through Jackson Arch.
The Commandant's office is located to the
right side of the sally port.

In 1988, VMI ceased requiring cadets to accept commissions in the military. Consequently, the percentage of cadets accepting commissions dropped to as low as 32 percent. Since General Peay's challenge to the Institute in 2003 to commission more cadets, VMI has actively encouraged cadets to accept commissions. Peay has worked hard with the VMI ROTC Command to ensure that VMI can offer as many ROTC scholarships as possible. At the 2008 graduation, 52 percent of those graduating accepted commissions in a branch of the service. The goal of the Institute is 70 percent. In addition, since 2002, over 63 VMI cadets have been called to active duty and deployed overseas by their National Guard and reserve units. More leave each semester.

The Virginia National Guard is commanded by Air Force Major General Robert B. Newman, (1973) a true example of a VMI citizen soldier. Newman holds a B.A. in Economics and was regimental commander and president of the Honor Court at VMI. After graduation, Newman entered active service and trained and served as an Air Force pilot. After leaving active duty, his jobs included serving as Deputy Director of Homeland Security in the National Guard Bureau in Washington D.C. and serving as Deputy Assistant to Virginia Governor Mark Warner for Commonwealth Preparedness. His service to the Institute continued through his illustrious career. From 2004-2006, he served as president of the Alumni Association Board of Directors.

General Peay in his remarks to the faculty and staff on August 22, 2003 addressed the role and responsibility of the military college in today's world. He stated:

> *There are two aspects ... to consider in seeking an answer [to the role of the military college]: one involves process and the other involves results. By process I am speaking of the way a military college is organized, its teaching methods, its use of discipline, its emphasis on efficiency and attention to detail, teamwork, character, personal honor, and the important place of physical and military training in fully achieving its special educational goals. By results, I refer to the successful formation, at the end of the academic course, of highly educated young men and women who have the knowledge and skills to be useful citizens and ready, if qualified to serve as citizen-soldiers.*
>
> *Our responsibility is to provide an intellectual and physical environment of the highest quality in which our cadets will develop and flourish as thinking and creative individuals, possessing high personal goals and values, committed to the service of others, and qualified to serve in the defense of our nation. At the center of this environment and foremost is the academic program, the intellectual development of cadets. Knowledge and the ability to think about complex issues and solve complex problems under pressure are the keys to success in peace and war.[7]*

Marine Corps Commissioning Ceremony
in the Cocke Hall courtyard.

Importance of Core Values

Entering the foyer of a large hospital in North Carolina, I was struck by a framed poster of the organization's core values. Value statements claimed, "We value honesty. We value personal caring. We value individual privacy," etc. Having previously participated in several exercises at another organization where we worked for months to define the values of its unique culture and organization, I wondered whether these values were really internalized by the employees. My experience told me that there would likely be a group of people who would violate these values on a regular basis; however, I guess when you display your core values in halls and elevators, there really is no excuse for not knowing them. But still, I am sure there are employees who never bother to read them. To them, they are just a piece of paper and some words.

As an American society, there are many core values we share. Many people fear that by allowing uncontrolled immigration our core values will change as people (sub-cultures) with different values migrate to our nation. Perhaps, but as a melting pot nation, we have always assimilated those populations who generally aspired to adopt our culture and the values that go with it.

So what do we as Americans value? Table 2.1 alphabetically lists the 15 values that U.S. adults of all religions, ages, genders, races, and social classes identify as the most important in our society.[8] When examining the table, one value stands out like a sore thumb: racism and group superiority. It is easy to explain group superiority. Everyone values being the best. VMI loves to think it is better than the Citadel and West Point loves to feel superior to the Naval Academy; but where does racism play into American values? How can we as a society possibly value racism? For sure, it is a contradictory value. After all, doesn't it contradict equality and freedom as treasured values?

As a white male raised primarily in the South, I have certainly witnessed racism growing up. Even today we all know it exists on every corner of suburbia and inner city street alike. I have also been on the receiving side of racism. Having lived in Hawaii, a state with less than a 20 percent Caucasian population, I quickly learned the locals had their own "N" word for whites. Young Hawaiians didn't like "haloes" dating their sisters or surfing "their" waves. In Vietnam, I discovered that Hispanics in my unit didn't care for African-Americans and vice-versa. Race riots and fights were common, and rarely reported in the newspapers back home. And to top it off, I discovered the Vietnamese, who worked alongside me, didn't like the three million Montagnard mountain tribesmen. These primitive people were not allowed to attend schools or to have books in their native tongue. At the time, they had no written language.

In 2003, we as Americans began to learn of and witness newer examples and forms of racism. Now we know the Kurds and Turks can't live together, not to mention the religious sects of Sunnis and Shiaa. Without question, VMI cadets and other prospective leaders need to understand the roots of racism, and why Americans and others in different cultures value something which publicly we say is bad. Like it or not, racism is an American core value, and a value shared and observed world-wide. Leaders must make informed decisions. Race will often be a factor in those decisions because we all live in a diverse society. Diversity has no place for racism. At VMI everyone works together in a diverse environment for the common good, regardless of race, gender, or religion.

Table 2.1
Core Values of U.S. Society

Achievement and success
Activity and work
Democracy
Education
Efficiency
Equality
Freedom
Humanitarianism
Individuality
Material comfort
Progress
Racism and group superiority
Religiosity
Romantic love and monogamy
Science and technology

Military Core Values

Another set of values to examine are the seven belonging to the U.S. Army. These Army values are believed by many to be the finest qualities a leader can hope to attain. VMI's Institution Outcome Goals require that every cadet understands "The values and ethical standards of the commissioned service to the Nation." For example, every member of the Army is educated in its values and every soldier carries a card in his or her wallet with the seven values … a constant reminder. Table 2.2 lists Army values as well as Air Force, Navy and Marine values, which reflect those things most important to members of the armed services, but also to society. By participating in ROTC, every cadet understands those values. These values when combined with respect and humility create effective units.

Table 2.2
U.S. Military Core Values

Army	Air Force	Marines	Navy
Loyalty	Integrity	Integrity	Honor
Duty	Excellence	Commitment	Commitment
Respect	Self Before Service	Sacrifice	Courage
Selfless Service		Perseverence	
Honor		Confidence	
Integrity		Integrity	
Personal Courage		Belief in Mission	
		Interdependence	

As a military college founded in Army traditions, it is appropriate for this book to examine the values of the Army. These values are the moral compass of the Army.[9] A soldier's duty is to assimilate these values and to act on them every day, in every situation. These values are not only the foundation cornerstone for developing our nation's military leaders, but leaders from all walks of life.

Honor. At VMI, the most cherished value is one's personal honor. The Honor Code requires students to be honest, fair, and just and to demonstrate regard for others' property. Honor and integrity are topics so important that Chapter 5 is devoted to a discussion of how these values can be internalized.

Respect. Cadets are taught respect for others, especially their faculty, classmates, and staff. Civility, compassion, and consideration of other points of view are all part of respect. VMI faculty and administration constantly remind cadets of the importance of being civil to others. On the rare occasion, when a cadet disagrees with me about a grade or an assignment, he or she can sometimes verge on being disrespectful. In the two cases, I can recall, the cadets later emailed me or personally apologized.

Duty. There can be no real leadership without duty. It is a leader's duty to accept responsibility for his or her actions and decisions. Leaders must ensure they have done all they can do as well as they can do it. This is also their duty.

Loyalty. Every VMI cadet is taught loyalty from day one. Without a doubt, the loyalty demonstrated throughout each Rat class is legendary. The relationships formed between Brother Rats ("BRs") lasts a lifetime. BRs are faithful, committed and devoted to each other. They form an incredible support group.

Selfless Service. The act(s) of providing assistance or support to others without regard to personal gain or profit is selfless service. A common example seen on campuses throughout the nation and at VMI is the many hours of service projects that students provide to a variety of non-profit organizations. Recent examples at VMI occurred when cadets participated in the construction of a Habitat for Humanity house and when they raised money for various projects to support their fellow cadets and others serving in Iraq and Afghanistan. Of course the ultimate act of selfless service belongs to the growing number of cadets who leave their classes to deploy with their Guard and reserve units to serve their nation and communities.

Courage. There are different types of courage. Some involve valor and bravery in combat, which VMI cadets demonstrated at the Battle of New Market. Others might involve such actions as reporting a BR for receiving unauthorized assistance on a paper. In the business world, it may be the CEO who looks in the mirror each morning during dark economic times and says "I will not be the one who lets this business fail." Courage is acting in accordance with an individual's beliefs in the presence of adversity, danger, or criticism.

Integrity. Honor, courage, respect, duty, loyalty, and selfless service are all wrapped around the seventh value of integrity. The action of making decisions based on shared values is integrity. In his leadership book, William Cohen identifies integrity as one of the eight universal laws of leadership, a law which is essential for all other laws to work. He defines absolute integrity as "doing the right thing."[10]

New Cadets sworn in at New Market Battlefield where VMI cadets fought and helped defeat Union Troops on May 15, 1864.
Source: VMI Communications & Marketing

VMI Core Values

In a recent welcome back speech, entitled "Bone Deep Caring," to the Corps of Cadets, VMI Superintendent General Peay laid out the importance of VMI values in a very inspirational and meaningful speech. He told our cadets:

> *"VMI is a community of learning, virtue, and discipline that prepares its cadets for the duties of responsible citizenship. Supporting our community are the following concepts:*

- *An educational program that mixes theoretical and applied knowledge a required core component, class attendance, progressive accomplishment over postponed perfection to produce useful citizens;*

- *A Spartan life in barracks, with a minimum of privacy, uniformity of dress, and adherence to a Code of Honor;*
- *Hierarchical forms of governance based on leadership;*
- *A military system of discipline designed to instill order, efficiency, accountability, respect for authority, and esprit;*
- *A meritocracy wherein all are forced to begin on an equal footing;*
- *A world in which the daily quota of time is short in comparison to the work to be done;*
- *A belief that development of character is best achieved through testing by adversity; and*
- *A system of physical development that develops a sound mind in a sound body.*

These beliefs and practices have distinguished VMI from other colleges for over 164 years, and they continue to guide us on a day-to-day basis. An important basic characteristic of a community is its tendency to develop shared basic assumptions about human relationships. These assumptions arise mainly in response to the community's need to develop "internal integration." But the process is not "automatic". Every group must learn to become a group, and every group must work to maintain and sustain itself as a group. What has been learned and created over many years can be unlearned and dissolved in a short time, if the community becomes complacent or turns down another path. In sum, another word for these shared basic assumptions is values.

Nothing reflects an institution more accurately than its values, those principles that hold it together, those principles for which it stands, those principles that are espoused by its members. VMI is a special college -- a tight-knit community of those who would learn and those who would teach and as such it is known across the land as an outstanding and rigorous academic community with high standards

*and commonality of purpose. But it is also known as
an institution with an especially strong commitment
to basic values.[11]"*

Support for VMI Values

A recent study surveyed senior executives who had
graduated from VMI.[12] Of this group, 53 executives responded
and ranked the importance of 30 values they believed successful
leaders should internalize. It should be no surprise that these
successful business executives ranked integrity/honor as number
one. This statistic is congruent with other national studies where
integrity and honor are also ranked at the top; however, there is
one difference: A higher percentage of VMI grads (92%) ranked
integrity/honor as their top value versus the findings of a national
study reported in the best selling *The Leadership Challenge* by J.
M. Kouzes and Barry Posner. The authors noted that 88 percent
of their respondents ranked integrity/honor as their number one
value.[13]

Other values ranked by the Kouzes and Posner were similar
to the VMI results, but there were a few major differences in
importance. For example, VMI executives ranked "dependability"
number two, where as Kouzes and Posner's research ranked it
number 11. As Rats, VMI cadets quickly learn they must depend
on others to survive. They clearly value those whom they can
depend on to see them through critical moments. Another key
difference was the degree of importance assigned to "loyalty."
VMI grads ranked it fourth, while Kouzes and Posner ranked
it as the 10th most important value. Perhaps, the Fourth Class
year again influenced the degree of importance placed on these
v⸏   by VMI grads. VMI Brother Rats are fiercely loyal to one

values not rated by the national study or ranked in its
⸏ "persistence" and "determination." These may be
⸏adets have internalized before entering VMI.
⸏s a few cadets, who in all due consideration,
⸏ changing colleges or at least changing

2

majors. These cadets possess tremendous determination in their quest to complete a VMI education. In many cases, this means multiple summer schools at VMI (we don't allow cadets in our major to take required courses outside VMI), a fifth year in the Corps, or perhaps marching endless tours for violating the *Blue Book* (book of regulations). Some of my cadets have marched as many as 94 penalty tours in a semester (100 tours and you are supposed to be kicked out). They refused to leave the Institute despite having no free time and being restricted to the barracks except for classes and official functions and duties.

A Culture of Success

VMI has a long and storied history, which frames the VMI experience for its thousands of successful graduates. While the Institute is quietly working to draw away from its close ties to the Confederacy, it is still believed by many Southerners that VMI is the West Point of the South, a claim also made by its archrival the Citadel. It is difficult to get away from that Confederate image when one instantly sees the Lee Chapel (Robert E. Lee's resting place) on the W & L campus near the entrance to the VMI Post. And of course, there are other reminders on Post such as Stonewall Jackson's statue and barracks arch or the "Virginia Mourning Her Dead" monument, where six of the ten cadets killed at New Market rest in front of the Nichols Engineering building. VMI administrators "are quick to explain that the Institute's homage to Jackson and the New Market cadets are not designed to celebrate the Confederacy but to recognize the values of honor, duty, and discipline embodied in praiseworthy individuals."[14]

My first year (2002) at VMI, reminded me that a small number of alumni still hold strongly to those old ties. While watching a parade before a football game, I noticed a huge Confederate flag being raised from the bed of a pickup truck behind the Corps and in front of Lejeune Hall (torn down in 2006 to build the new barracks). I think most spectators and everyone involved with VMI was embarrassed. The individual responsible was quickly asked to remove his truck and flag from

Post. While VMI may have been the West Point of the South, many of its initial leaders and faculty including General Jackson were graduates or had close associations with the U.S. Military Academy. What most of the public doesn't realize is that 15 (12 as officers) VMI cadets fought for the Union in the Civil War.

While VMI will always be associated with the history of the Civil War, it would prefer to be known as an outstanding academic institution that produces citizen-soldiers. I believe VMI is well on its way to developing an image of sustained academic excellence. From 2002 to 2007, VMI was rated by U.S. News & World Report as the best Public Liberal Arts College in America category. In 2007, the Carnegie Foundation categories changed and VMI was placed in the company of the U.S. service academies. In the latest ranking, VMI was number three, alongside number one Naval Academy and number two West Point. VMI views this as pretty good company to be grouped with. While this ranking of public liberal arts institutions does not include the private Ivy League colleges or large, research universities, it does include all public colleges where undergraduate education and teaching are the focus of their mission.

Recruiting. To achieve an organization's mission, it is always easier to select people who already have the values you seek or at least they aspire to achieve them. This applies to corporations as well as colleges. Just as George Marshall carefully selected his staff and commanding generals in World War II, so does the admissions staff at VMI carefully select each year's class of Rats. Colonel Vernon Beitzel, the long-time director of admissions, and his staff are trained to look for young people with the "right stuff" to succeed at VMI. A sample of high school extra-curricular activities is well represented in Table 2.3.

Statue of General Thomas "Stonewall" Jackson
alongside Matthew, Mark, Luke and John cannons
from the Civil War era in front of the "New" Barracks."

Table 2.3
Extra Curricular Profile of the Class of 2010

Officer, student body/class	13%
Team captain	42%
Student government	13%
National Honor Society	25%
Athletic team (2 years)	84%
Boys'/Girls' State	12%
Eagle Scout/Gold Award	13%
Club Officer	25%
JROTC or CAP	25%
Band	18%
Chorus	7%

On the flip side, the academic dean and Institute department heads are constantly evaluating faculty applicants for their interest in working closely with cadets and being able to adapt to a military environment and traditions. During my tenure at VMI, I have seen a few odd duck candidates for faculty positions, who clearly would not have fit into our culture of cadets first. Fortunately, the faculty selected understands what our cadets need and expect ... excellent and caring teachers.

Admission in the early years of the Institute was based primarily on "character and stability." Academics or intelligence beyond average was not a piece of the admissions selection process. The Board of Visitors (Institute's governing body) recommended one student per year per state senate district. These students had to be residents of Virginia. They received free tuition, room, and board and in turn were expected to serve in the state militia. This system worked for the first 20 years and the enrollment grew to 300 by 1860. The first out-of-state cadets were admitted in 1857 due to vacancies caused by a financial recession. The image of VMI as being a place for the average student was probably strengthened by George Marshall who was a very weak student prior to admittance to the Institute. Even his brother, Stuart, did not want him to attend VMI thinking that George would "embarrass the family." Marshall was admitted because of his family name and character and because his brother was a successful graduate.[15]

Both the former dean of the Institute, Brigadier General "Casey" Brower and Lieutenant General Josiah Bunting, VMI's 13th Superintendent and a Rhodes Scholar, take issue that the original mission of VMI was to accept average students. Bunting does not believe that the initial mission is relevant to the 21st century.[16] He has said:

> *"It is a conviction very strong in me: now that we do ourselves little good, and that we really do not understand what VMI is about when we say our mission is to find and admit only average young people and help them attain the best achievements of which they are capable. In fact our mission is to look beyond*

our culture's simple-minded idea of what constitutes true potential and promise. There is of course no reason to assume that these things [standardized tests such as the SAT and ACT] do not mean their holder is not able, and promising: but there is no reason to assume that he will be the most able, the best suited, the most likely, to succeed in the world's fight. I do mean potential and fitness for leadership in our democratic republic: in the institutions of government and military service, in the law and business, in the fields of human endeavor in a great commercial republic."

Later in his address Bunting proclaimed:

"VMI's business must be to find and enroll and educate such men and women as these: holding in death's grip our time proved irreducible of our heritage as honorable citizen-soldiers, ready respond in times of deepest peril: but equally facing our future with delight and confidence, not fear or resentment.

Our course is now set firmly, its objectives point no different from its destination 160 years ago: it is the education of young cadets who will understand that leadership is their obligation, as it is their heritage, and who will aspire, by emulation, to be the leaders of the 21ˢᵗ Century and citizens their forefathers were in the 19ᵗʰ. Men and women of singular eminence of mind and character: not of character without cultivated, self-educating intellect; not of mind not rooted in the firmest elements of character (self-reliance, resolution, and selflessness). Perhaps the world outside does not think of a military college as a house of intellect and originality, but I argue that it can be the best house of mind of all as it is the best crucible for the foraging of character: a place in which true eccentricity will flourish, and the false allowed no place."

27

A Typical Rat Class. In August of 2008, the Class of 2012, 446 strong, matriculated to VMI. Although many more out-of-state students applied for admission than in-state students, the class was composed of 56 percent Virginians. Despite public opinion, this class has only five percent of cadets whose fathers attended VMI. Fifteen percent will have had at least one relative who attended. While the number of legacy students is declining, it is still not unusual to find cadets who are part of three and four generations of grads or who have had several brothers or sisters attend the Institute. In 2008, one of my cadets, Cam Hagan, was a fourth generation cadet. His uncle, grandfather and great-grandfather were all VMI graduates. His cousin, Addison Hagan, was president of the class of 1997 as was his father (1968). Cam admitted there was "just a little pressure" to attend VMI.

Cadet William "Cam" Hagan III represents four generations of Hagan alumni. His cousin, Addison Hagan, and Cam's uncle were both presidents of their classes.

With Virginia Tech and Mary Baldwin having a corps of cadets as well, it is much more difficult for VMI to recruit qualified female cadets. Five percent of those who matriculate will be female. The goal is to enroll 15 percent of each class with females, a number which approximates the number at the service academies and armed services. In 2008-2009, 39 women (8.7%) matriculated to VMI.

In 2007-2008, VMI hit the jackpot. The Institute enrolled its first known set of triplets, two brothers and a sister. All survived the Ratline and are doing well in different majors ranging from International Studies and Psychology to Mechanical Engineering. The Redmond triplets, Angela, Stephen, and Thomas, are assigned to different companies and undertaking different activities. Angela, the original holdout, was convinced by her brothers and coaches to play soccer for the women's team. A former team captain and goalie on the state championship team from Woodson High School in Fairfax, Virginia, Angela earned All-Freshman Team honors from the Big South Conference for her play as goalie. Stephen and Thomas are both pursuing commissions as Army officers. Thomas has contracted with the U.S. Army and is on an Army scholarship.[17]

American Caucasian students made up 85.6 percent of those enrolling in 2007, African-Americans 5.7 percent, Asians 3.4 percent, and Hispanics 5.7 percent. Less than one percent are international students, but they come from a wide variety (9) of nations. I taught cadets from Mexico, Russia, Thailand, Korea and the Republic of China. Over the years, some of the highest ranking military officers from Thailand, the Philippines and the Republic of China have been VMI graduates.

The great majority (77%) of enrollees graduated from a public high school and six percent attended another college before enrolling at VMI. I had one such young man, "Steve," who had attended three years of college at several schools before matriculating to VMI. He told me he had an awaking one day and decided he wanted to prove to himself and to his family that he was VMI material. Thus, although technically a junior (Second Class cadet) based on his credit hours, he entered as a 22-year-old Rat. VMI was a little more of a challenge than he anticipated, but he did finish with an additional year. Once again determination and persistence as a goal rear their heads and prove to be major driving forces in earning the coveted VMI ring and diploma.

General Bunting is correct about his academic assumptions. Certainly today's cadets are not average. Their high school grade point average is a B (3.3-3.5 median range) and their composite SAT score (math/verbal) is 1141 (ACT composite is 24). This score is more than 100 points higher than the national SAT average and that of a major VMI competitor (non-service academy). Having worked for eight years as a MALO (Military Academy Liaison Officer) for West Point and as the faculty chair of the admissions committee at a liberal arts college, I know that many admissions officials believe that there is a measurable difference in student performance for every 50 points in SAT scores. Thus, the middle SAT range (1060-1200) provides VMI with cadet classes that are more than capable of holding their own compared to the average student attending college today. While it should be stressed that SAT scores alone are poor predictors for success at VMI or at any college, intelligence is a major factor in developing the competence needed to be a leader. The SAT/ACT, however, cannot predict the determination or persistence with which a student will accept the mental and physical challenges of VMI. SAT and ACT scores are only one indicator that a student has the ability to succeed academically. I am sure cadet records will reveal that many cadets with high SAT scores did not graduate from the Institute, and on the other hand, I am also sure there are just as many former cadets with below average SAT scores who graduated and were very successful in their careers and life in general.

Internalizing Values

While Americans find it difficult to accept a total culture concept, such cultures do exist in countries such as Japan. The Japanese have traditionally identified with their employer and adopted their values. In Japan some employers provide housing in company apartment buildings. If I asked Japanese workers what they do for a living, a common response might be "I work for Toyota," or whichever company employs them. In the U.S., our military is the closest to a total culture that one can find. Families are closely bonded through deployment support groups, family housing, recreation facilities, chapels, and shopping on post. This is especially true for families stationed in Germany, Italy, and Korea among other global assignments. Even within the military, each branch has its own culture and set of values. Often this culture and its values stick with a person forever.

Because the VMI culture is created by a variety of intense common experiences, VMI values tend to continue into business and family life. When VMI was going through the experience of first fighting the admission of women and then enrolling them, a bumper sticker was commonly seen on cars, which read, "If you want the VMI experience, marry a grad." I am sure my wife can identify with that bumper sticker. As a Virginia Tech Corps of Cadets alumnus, I highly value organization in my life. I still fold my towels the way I was taught as a rat and organize my closet and drawers pretty much the same way as 40 years ago. I also believe strongly in the importance of good character, honesty and integrity ... all Virginia Tech Corps values. However, in my opinion, the key difference in my Virginia Tech experience and that of VMI alumni in the same period (mid-late 1960s) is the way the VMI culture gels and cements each class and then maintains contact with alums through the class agent system. Each class agent is responsible for collecting news on alumni and keeping alumni connected to the Institute. This, I believe, is an important reason why VMI has the highest per capita ($380 million endowment as of 2007) giving of any public college or university in America.

The Army uses five steps to achieve the internalization of Army values.[18] A review of these five steps indicates that VMI uses a similar model to internalize core values within our cadets. The first step is the self-identification and selection process. This has been previously discussed in this chapter, but there are scores of cadets who after 9/11 decided to select a military college, and for a variety of reasons ended up at VMI. The second step involves an early socialization process. VMI does this through an intense matriculation process one week before classes start in la
For some, it starts earlier as they participate in the

31

Transition Program (STP). This four-week program educates new students about daily cadet living, gets them in shape, and allows them a chance to ease into academic work without the pressures of the "Ratline." The third step is to provide role models. Approximately 17 percent of the faculty are military veterans. The Army faculty and staff veterans wear the black beret outside of class. In addition, an outstanding ROTC cadre teaches not only military subjects, but also act as Officers in Charge of Barracks on a rotating basis. Most of these personnel are veterans of the Global War on Terrorism.

Step four is sharing stories and examples. It is extremely important that stories both positive and negative be shared with cadets. Who was successful and why? Who failed and why? It does no good to cover up the truth or to perpetuate war stories that never occurred. Unfortunately, this is what happened when Army Sergeant Pat Tilman, former NFL star, was killed by friendly fire in Afghanistan. Cadets need to know what happened, and why, so it won't happen on their watch. During General Peay's "Bone Deep Caring" speech, he shared personal stories as a cadet and an officer to illustrate how his VMI values were put into action. Finally, the Army uses feedback and performance evaluations to monitor and reinforce the internalization process. It is paramount that assessment be built into a leadership development system. VMI is currently in the process of developing a more comprehensive process for assessing cadet leadership development.

Conclusion

VMI and its cadets, alumni, and faculty and staff stand on high ground. That high ground symbolizes a foundation of high standards, which consist of values and principles that every cadet must meet to graduate and aspire to maintain for a lifetime. VMI is a place where citizen-soldiers like George Marshall and many other great leaders have served their states and nation. It is a place where the little things matter; where cadets are taught through a rigorous academic education and challenging military and physical experiences that it is better to take the harder route than the easier paths through life. At VMI honor, integrity, duty, loyalty, selfless service, and courage count. Through a four-year process cadets are taught to internalize these values until they become a part of them, not just a uniform to put on and take off as the environment of their daily lives changes.

LEARNING TO LEAD
A WHOLE PERSON PROCESS

"Fit no stereotypes. Don't chase the latest management fads. The situation dictates which approach best accomplishes the team's mission."

**Colin Powell, Army General
and Secretary of State**

Organizations spend tens of millions of dollars annually on leadership training. It is estimated that 42 percent of leaders spend up to five days attending leadership development programs each year. Another 28 percent spend more than five days annually attending leadership development programs. The astonishing result is that only 10 to 15 percent of these training programs result in sustained change. In other words most people keep doing the same old things day after day.

An old Buddhist saying holds that "to know and not do is not to know at all." Thus, much of this leadership training is proving to be ineffective, yet research demonstrates that leadership effectiveness accounts for 20-45 percent of organizational effectiveness. Effective leadership training must be more than learning facts. It must develop the whole person. Aristotle may have said it best when he stated "we are the sum of our behaviors – excellence therefore is not a single act but a habit."[1] And thus it is at VMI, where cadets learn to lead from a four year leadership development process focused on developing those values and characteristics deemed important in the whole person. This chapter examines various methods used to develop leaders.

Learning from Leaders

Leaders can be categorized as teachers, heroes, and rulers.[2] To help bridge the gap between the science of leadership and the art of leadership, let's examine personal glimpses of different types of leaders at different levels, and how they led during leadership moments. These personal glimpses illustrate that leaders do not look or act in the same ways.

Teachers. Teachers are said to be "rule breakers" and "value creators."[3] A few examples from history include Jesus, Plato, Socrates, Aristotle, and Aquinas. VMI is blessed with many great teachers. It has the best teaching faculty that I have witnessed in my 33-year career in academics. My experiences include serving at a large state university, a service academy, a small liberal arts college, and finally a state military college. The VMI faculty works as a team for the benefit of their students. They are truly dedicated and fully committed to helping students reach the high standards necessary for their future success. This, in my experience, is fairly unusual and a main reason why I chose to join the faculty at VMI.

I first realized how serious VMI was about faculty leading by example when, in mid-August of 2002, I found myself along with my assigned mentor, Colonel Francis Bush, and approximately 25 other faculty at Eagle's Landing. This rustic retreat is an outdoor adventure camp bordering the Allegany Mountains near Fincastle, Virginia. As a "Rat" faculty member, I was asked to participate in a very challenging and physical three-day boot camp of sorts that was designed to integrate new faculty into the VMI team, bond us together, and to orient us to Institute policies, procedures, and other important information. Many of the new faculty were young and in their early to mid-30s. I was weighing in at 57 years old, carrying 15 extra pounds, and returning from a hospital stay for a kidney stone the week before camp.

You get the picture. We started slow with some low-ropes, low-impact activities and then built up to the, "Oh, no! Do I have to do this?" One of the first activities was the 40-foot climbing wall. My mentor, who had apparently done this many times before, tried to show us how easy it was by doing it twice, the second time blind folded. I was impressed and managed to slowly make my way up the incline cheered on by two young college-age girls who were instructors at the course. How could I possibly fail?

My next big challenge was really BIG! We were told Eagle's Landing had the second longest zip wire (approximately 800 feet) in the world. Our group sent faculty up the mountain one by one, while the rest of us watched as our colleagues jumped off a platform and headed straight down the mountain hanging from a harness attached to a cable or zip line. Most figured out how to slow their speed by thrusting their arms and legs out to act as drag; however, our new Air Force professor of air science, Colonel Bob Lind, tucked into a ball and headed full speed down the mountain to the applause of the group. Okay, I thought, Bob is a brave guy and probably half crazy, and compared to bailing out of an F-16 at 400 miles per hour, this must seem like a piece of cake. Finally, I decided I couldn't be branded a wimp. I headed up the mountain, but not quite psyched enough to yell "airborne" and jump. Summing up all my courage, I strapped in and finally made the leap for life that barreled me down the mountain with my arms and legs spread to the max. What a sight!

And so is the spirit of the VMI faculty. They don't just sponsor the club hockey team, they play on the team with the cadets; they don't just sponsor the marathon team and accompany cadets to NYC and Washington for races, they run the races with them; they don't tell cadets to climb House Mountain, they hike the mountain with them and join the Rats in completing phases of "Rat Challenge;" they don't just tell the pep band to show up for the basketball game, they play the trumpet with them; nor do they tell cadets about the Battle of Iwo Jima or the New York Stock Exchange, they take them to those places for once-in-a-lifetime visits. Thus, after spending my first year at VMI as the Wachtmeister Visiting Chair in Science and Technology, I understood the value of the Eagle's Landing experience. At VMI, I was expected to lead and develop cadets by example. Many of these leadership experiences would take place outside of the classroom.

There are many well know faculty associated with VMI such as General Thomas "Stonewall" Jackson (civil war leader and West Point graduate), Commodore Matthew Fontaine Maury (founder of the Naval Observatory and noted oceanographer),

Captain John M. Brooke (inventor of the Brooke smoothbore naval guns) and General G.W. Custis Lee (R.E. Lee's son and later president of W&L).

A lesser known faculty leader is Colonel Floyd Duncan, chair of the Economics and Business Department and a VMI (1964) chemistry graduate. Duncan served as an officer in Vietnam and was awarded the Combat Infantryman's Badge and the Bronze Star. After earning an MBA in 1968 and a Ph.D. in Economics from the University of South Carolina in 1972, he joined the VMI faculty as a member of the Economics department in 1978. During this time, he served in the U.S. Army Reserve and eventually rose to the rank of Colonel. He is a 1985 graduate of the resident Army War College course and served as the Deputy Commander for Logistics for the 310th Theater Army Area Command. While remaining active in the reserves, Duncan's career also flourished at VMI. He rose to the rank of professor and served as department head in the early 1990's. In 1995, he received the coveted and much honored VMI Distinguished Teaching Award.

Colonel Duncan's accomplishments certainly demonstrate leadership; however, none so much as an action I witnessed in the spring of 2005. Floyd and I were talking outside our offices in Scott Shipp Hall, waiting for a departmental faculty meeting. The topic of conversation was the near completion of his much anticipated retirement home on Smith Mountain Lake in Virginia. Floyd was very excited about the prospects of his near retirement and moving into a new home in just a matter of weeks. However, in an instant, everything changed. At the departmental meeting our chair, Colonel Ed Sexton, announced that he had taken a position at BYU Boise in order to be closer to his family out west.

I knew Duncan was the most senior member of the faculty and would likely receive the support of the department to replace Colonel Sexton, but after our prior conversation moments earlier, I figured there was no way he'd stay longer to complete our challenging five year process to achieve AACSB International accreditation. Since we were just finishing our first of five anticipated years, Colonel Duncan would have to commit to an extension well beyond his intended stay at VMI. Much to my surprise, he seized the moment and accepted the position, when asked by Brigadier General "Casey" Brower, Academic Dean of the College.

Duncan later told us at our first faculty meeting that he would "stay the course" until accreditation was earned. Given that we are a department of 14 faculty members and not a school or college of business, this is a very formable task. When hopefully completing the process in October of 2009, VMI likely will be the smallest of approximately 550 business programs accredited throughout the world. Most accredited business programs are larger than VMI as a whole. There is no doubt that Colonel Duncan's leadership will have added value to the cadet experience and have a lasting impact on VMI and its students.

VMI has also produced another "teacher" leader, Thomas Morris (1966), the Secretary of Education in Virginia in Governor Tim Kaine's administration. Prior to his appointment, Morris was the popular president of Emory & Henry College for 13 1/2 years. A well-known Constitutional scholar and political scientist, Morris taught at the University of Richmond, where he was the University Distinguished Educator and professor at the University of Richmond for 21 years. Dr. Morris authored or co-authored four books, which include one book written with popular political analyst (national TV/radio), Larry Sabato of the University of Virginia.

Heroes. Heroes have a different style than teacher leaders. They are responsible for "great causes and noble works."[4] Historical heroes include such people as Edison, Pasteur, Einstein, Galileo, and Michelangelo. VMI has been associated with many great hero leaders, including General J. H. Binford Peay III '62, former CENTCOM commander and 14th superintendent of VMI, who is the visionary, architect, and renovator of a reconstructed and modern VMI Post (campus).

Of VMI's many heroes, one should also never forget the ultimate sacrifice of class of 1962 valedictorian and seminarian Jonathan Daniels' life at Hayneville, Alabama to the civil rights cause. Dr. Martin Luther King said of Daniels' actions, "One of the most heroic Christian deeds of which I've heard in my entire ministry and career was performed by Jonathan Daniels. Certainly there are no incidences in the annals of church history, and though we are grieved at this time, our grief should give way to a sense of Christian honor and mobility."[5] More about Daniels' courageous actions can be found in Chapter 4.

And of course no leader was more accomplished than George C. Marshall (1902). My desire to know more about this hero attracted me to VMI. Once at the Institute, I began writing a leadership book about General Marshall. The book, *George C. Marshall: Rubrics of Leadership*, was published in 2006 by the Army War College Foundation Press. After reading Marshall's personal papers and the volumes of books on him, I quickly understood why so many people admire this great American hero. After leading the U.S. war effort as army chief of staff in World War II, his most significant and lasting act was probably the conception of the Marshall Plan, for which he fought so hard to earn popular and Congressional approval. His humanitarian efforts to rebuild Europe and to stop the spread of communism eventually won him the Nobel Prize for Peace. Today a statue and the New Barrack's arch at VMI stand in his honor. Marshall, who passed away in 1959, was a man for all times.

Rulers. Rulers are leaders who are chiefly motivated to "dominate others and exercise power."[6] Consider how Alexander, Julius Caesar, Napoleon, Elizabeth II, Mao Tse-Tung, Lee, and Washington led their troops. Since most rulers are military leaders, it is only fitting that VMI produce or be associated with great military leaders (approximately 18% of graduates choose military careers). Some easily recognizable leaders associated with the Institute include Lieutenant General George Patton Jr. (1908), Lieutenant General John A. Lejeune, Rear Admiral Richard E. Byrd (1908), Lieutenant General "Chesty" Puller (1921), and General John Jumper (1966). While Patton graduated from West Point, he spent his Rat year at VMI and was one of a long line of Pattons to attend and teach at VMI. Patton declared as a youth he wanted to be a hero. He is an excellent example of a ruler leader. His exploits and victories in Europe and Northern Africa in World War II are legendary. His final combat command was the Third Army, which broke out of Normandy and pushed into central Germany and northern Bavaria. By V-Day Patton had reached Lenz, Austria to defeat the Germans.

Learning from Books and School

The third most common source cited by leaders is learning their trade by reading books and taking classes.[7] A quick glance at the shelves of any Barnes & Noble book store would make it obvious that there are an abundance of leadership books from GE's Jack Welch to Duke's Coach Mike Krzyzewski and hundreds of other leadership gurus. Someone must buy these books or publishers would not continue to grind out scores of new titles each year.

College courses are another source of knowledge. For example, I teach a course titled "Business Leadership & the Classics." Cadets in my course study leadership through a series of case studies using characters from the classics of literature and film. MBA programs are also putting an increasing emphasis on leadership courses, and in the 1990s the University of Richmond created the Jepson School of Leadership, the first leadership school of its kind, which awards an interdisciplinary undergraduate leadership degree. VMI currently offers a leadership minor and employs three full-time professors to teach and research leadership. These professors are responsible for initiating leadership programs through VMI's new Leadership and Ethics Center.

Other universities, corporations, associations, and nonprofits offer thousands of classes each year. Corporate executives can go off for weeks at a time to executive education seminars at Columbia, the University of Virginia, Wharton, Duke, Stanford, and Harvard to name but a few of the better known programs. Even the federal government operates the Federal Executives Institute, which provides leadership courses for senior level government leaders from over 200 federal agencies.

Table 3.1 VMI Leadership Development Model: Followership to Leadership[8]

Intellectual	Military	Athletic	Ethics and Moral	Leadership
Core Curriculum	Regimental System	Varsity Sports	Honor System	Required Leadership Course
Majors	Class System	Intramural Sports	Honor Court	Corps Leadership Positions
Minors	Mandatory ROTC	Club Sports	Honor Education	Class Officers
Concentrations	Rat Line	Physical Training	Community Service Projects	Team Captains
Honors Program	Dyke System	Rat Challenge	Chaplain's Programs and Religious Clubs	Honor Court Officer
Writing Intensive Courses	New Market Ceremonies/ Breakout	Required PE Courses	Post Service Projects	Club Officers
Internships	Field Training Exercises	VMI Fitness Test		Rat Challenge Cadre
Study Abroad	Summer Camp Training			Hell Week Cadre
Foreign Student Exchanges	National Guard and Army, Marine, and Air Force Reserves			ROTC Leadership Courses
Undergraduate Research Initiative				Internship Project Leadership
Peer Tutoring				Acad. Tutor
Jackson-Hope Fund				First Class Cadet Dyke
Academic Clubs				Class Project Team Leaders

Are Leaders Always Successful?

There is growing evidence to support that often a leader's success or lack of success is due to his or her environment and circumstance. For instance, I once worked for a college president, who was very successful in his previous job as a president of a much larger state university. His entire life had been successful, and he believed that trend would continue. He was quite intelligent and felt he had a leadership formula that would work anywhere with anybody. After all he had previously moved through faculty ranks and administrative roles without missing a beat. Initially, he was the darling of our campus and community circles. His self-confidence was evident, but as time progressed his ego and sometimes condescending attitude became all too well known across campus. In less than four years, the campus community turned on him, and nothing he could do (right or wrong) seemed to win faculty approval.

His direct style of leadership, while perhaps appropriate and successful at larger state universities, did not match the requirements for success at a smaller college where gossip and academic self-governance ruled. The campus cultures were so very different. After four years, our president appeared depressed, somewhat withdrawn, and he failed to make some critical decisions, possibly to avoid increasingly stressful situations. Finally, his health got the best of him and he resigned. This example illustrates the importance of social interactions as a component of leadership. You just can't place a George Patton in the position of Secretary General of the United Nations.

Before accepting a leadership role, leaders must carefully analyze the environment in which they will be working. It may be impossible to succeed. Such was the case of George Marshall in China after World War II. Marshall tried to unite the Nationalist and Communist parties to sustain one China, but his tireless efforts were not successful.

Another illustration involves the integration of women into the Corps of Cadets at VMI. This was no easy feat, but after the Citadel's less than flawless attempt, VMI administrators were determined to make an honest attempt to do the right and best thing for all cadets. It should also be noted that after a battery of law suits to remain single-sex, the Supreme Court mandated in 1996 that VMI become a coeducational institution. The cost of VMI's efforts came to seven years and approximately $10 million (spent from the VMI Foundation endowment).[9] The Institute's Alumni Association and the VMI Foundation even went as far as to consider closing down or privatizing the school to avoid admitting women. The final vote of the Board of Visitors was 9-8 in favor of coeducation.

It was hardly a positive environment in 1997 for the new female Assistant Commandant Sherrise Powers. Major Powers was well-qualified for her new position, although being the only female allowed to walk the barracks stoops offered a huge challenge. Despite a law degree and eight years of prior active duty service as an Army noncommissioned officer and nine years in the Army reserves, she was met with extreme bitterness and hostility by many male cadets. Her role as a leader of all cadets, male and female, was constantly challenged. She was the first female officer with the power to give demerits and make corrections on the spot. In one incident, it is reported that hundreds of cadets leaning over the barracks' railings called her insulting names and told her to leave after she broke up a ball game in the central courtyard.[10] While Powers handled the situation the best she could, once again, the environment and circumstances dictated her effectiveness in a leadership moment. In the Army, this incident would never have occurred, and if it had the violators would have been severely punished.

Earlier times at military colleges and academies would suggest that failure was not an acceptable option. West Point cadets learn there are only four responses to an upperclassman or officer: "yes sir," "no sir," "no excuse sir," and "I don't understand, sir." Today, VMI administrators will tell you they expect cadets to fail and that the system encourages failure so cadets can learn from their mistakes. Cadets must learn that everyone makes mistakes. Bobby Jones, the great golfer of the 1920s, once said after losing a tournament by one stroke, "I never learned anything from winning a tournament."[11]

As individuals, we must fess up to our mistakes in order to improve our performance.

I learned a valuable lesson as a business school dean, when I misread the campus political climate and failed to persuade a new administration of the importance of AACSB International accreditation. If I had successfully convinced our new president to drop our less than thriving MBA program until we could reassess our market and build a new program, we could have achieved accreditation at the undergraduate level. Looking at campus leadership from a fundraising side, the president was afraid of alienating our MBA alumni. We got very close to achieving our goal of accreditation, but close wasn't good enough. Thankfully, VMI understands the importance of AACSB International accreditation.

Reflecting on this leadership moment, I now realize now that I failed at performing two key leadership skills: persuasion and influencing. Consequently, I decided to put my ego aside and step down after eight years of leading our school. It was clear to me that accreditation wasn't going to happen on my watch. A lot of personal and sometimes painful reflection was spent evaluating my decision to retire early (I was tenured) from the college, but I knew I needed to improve my skills and to learn from my mistakes. Working in the same environment would not have provided me with the new challenges I needed. I decided it was time to move on. I then turned my attention to training my successor. In my business leadership class, I talk openly about both my successes and mistakes as a leader. I emphasize ways that students can avoid similar mistakes such as not correctly understanding the culture and the power of office politics.

In 2007, a committee of cadets, staff and faculty rewrote the cadet *Blue Book* or cadet regulation book. The new *Blue Book* is far less prescriptive and encourages cadets to make decisions based on what is right versus doing something because the *Blue Book* says one can or cannot. This paradigm shift occurring now will require a major change in individual attitudes and Institute culture. It will not be achieved overnight and in all likelihood won't be fully accepted before 2012 (period of time when five classes will graduate that were under the old *Blue Book*).

Other modifications towards a more inclusive system of leader development are also being considered by the Institute so that graduates can adapt more quickly to fast changing situations in the board room and battlefield. The eventual product will directly tie cadet development goals (found in the Institute's Strategic Plan Outcome Goals) to specific leadership experiences. Cadets will be expected to demonstrate the achievement of the following goals by:

- The ability to anticipate and respond effectively to the uncertainties of a complex and changing world;
- Intellectual curiosity, imagination, and creativity;
- The ability to recognize moral issues and apply ethical considerations in decision-making;
- The ability to act rationally and decisively under pressure;
- Mastery of basic military skills required for entry into the commissioned services;
- A commitment to physical fitness and wellness, including the physical skills required for entry into the commissioned service; and
- The ability to understand and apply the art and science of leadership to inspire, motivate, and develop subordinates, accomplish organizational goals, and lead in a complex and changing world.

Conclusion

There are many sources from which one can learn leadership skills. Some individuals learn by observing other leaders, while some may learn from books, college courses or seminars. Effective leadership training will develop the whole person, and thus encompass the intellectual, military, athletic, ethical, and moral aspects of the individual. A leader's education will include discussions on different types of leaders, their styles, and their successes and failures. Most will fit the profiles of teachers, heroes, or rulers. All will need to think under pressure and adapt to a constantly changing and complex political, economic, and physical environment. Our rapidly changing times will require leaders who can think on their feet and can offer creative and imaginative solutions to a multitude of problems. It will also be imperative that our future leaders make decisions within an ethical and legal framework.

The remaining chapters will examine the "black box" of leader development and demonstrate how a broader definition of leadership development fits the requirements of the 21st century leader.

LEADING FROM THE TRENCHES
CHARACTERISTICS OF VMI'S AUTHENTIC LEADERS

"From the first day as a Rat to that proud day when we receive a VMI diploma we are taught that honesty, self-discipline, teamwork, sensitivity to others, personal acceptance of responsibility and accountability, individual effort, courage and the brotherhood of man are the foundations of a successful democratic society."

John D. deButts 1936,
former Chairman of the Board, AT&T

In the early 1990's the Justice Department began a series of trials before Federal courts aimed at requiring VMI to admit women. The government's chief educational expert was Dr. Clifford F. Conrad. He testified that the Institute offered a unique educational experience that was built on five components: education, military training, mental and physical discipline, character, and leadership. Conrad stated in testimony that eight interlocking systems held the VMI experience together and made it different from the ones offered by cadets enrolled in the Virginia Tech Corps of Cadets. The eight systems included a rigorous academic system, the military system, the honor system, the class system, the dyke system, the Rat system, athletics, and the Spartan barracks culture. While the Institute disagreed with Conrad's conclusions about the integration of women into the Corps, he was right on in describing the VMI experience.

In previous chapters, we have defined and reviewed VMI's unique leadership development model. This five-component model for leader development is today anchored by the new (2008) Center for Leadership and Ethics, which was created to support the critical mission of developing citizen-soldiers to lead our nation. One such citizen-soldier was 1st Lieutenant Terry L.

Plunck (1988). He was one of 522 VMI alumni to serve in either Operation Desert Shield or Operation Desert Storm. Lieutenant Plunck was a member of the XVIII Airborne Corps' successful effort in 1991 to drive Iraqi troops out of Kuwait. Plunck, who received his commission from Major General Binford Peay, commander of the 101st Airborne Division during Operation Desert Storm, died in action on February 26, 1991. His unit was clearing mines around the Kuwait City airport in advance of U.S. troop movements. The Vinton, Virginia native was an exceptional cadet. Plunck was a cadet captain (S-2), active in the Fellowship of Christian Athletes (FCA), and was the recipient of the prestigious Cincinnati Medal. This award is given annually by the faculty to the cadet who distinguishes him/herself by efficacy of service and character throughout their cadetship. Certainly service and character are characteristics of authentic leaders. This chapter discusses these and other characteristics sought and admired by those who identify present and future leaders.

General Binford Peay III, VMI superintendant,
and Donald Rumsfeld, Secretary of Defense,
pause to chat before the 2006 graduation ceremony.

Leader Characteristics

Once cadets complete the leader development program and graduate from VMI, a fair question to ask is, "What does the ideal VMI product (graduate) look like, or what are the characteristics of any ideal leader?" In their book, *Credibility,* James M. Kouzes and Barry Z. Posner identified through extensive research 10 key characteristics of "admired" leaders. In order of importance admired leaders are 1) honest, 2) forward-looking, 3) inspiring, 4) competent, 5) fair-minded, 6) supportive, 7) broad-minded, 8) intelligent, 9) straightforward, and 10) courageous.[1]

In another best-selling leadership book, *Lessons from the Top,* Thomas J. Neff and James M. Citrin, after conducting an extensive analytic research project, identified America's top 50 (as opposed to all) business leaders.[2] Their choices included such well-known leaders as Steve Case (AOL), John Chambers (Cisco Systems), Michael Dell (Dell Computer), Elizabeth Dole (Red Cross), Bill Gates (Microsoft), Andy Grove (Intel), Herb Kelleher (Southwest Airlines), Howard Schultz (Starbucks), Charles Schwab (Charles Schwab), Bill Marriott (Marriott International), Fred Smith (FDX), and Jack Welch (GE).

The authors were able to further list the top 10 traits that their best leaders had in common. The top ten traits included 1) passion, 2) intelligence and clarity of thinking, 3) great communications skills, 4) high energy levels, 5) egos in check, 6) inner peace, 7) capitalizing on early life experiences, 8) strong family lives, 9) positive attitude, and 10) focusing on doing the right things right.

It should be noted that while VMI can't create strong family lives as cited in the Neff and Citrin study, there is substantial anecdotal evidence that most cadets come to VMI with strong family ties, which influence their decisions; and it is their family values and ties that enable many cadets to survive and thrive from a rigorous Institute life. To illustrate the point, the example of one graduate is offered whose family values affected both his decision to attend VMI and his performance and attitude in developing into a superior cadet and Army officer.

In the fall of 2007, I was asked by a non-profit organization to assist in prepping a former cadet for an interview at the Harvard Business School. The next week a group of four of us met at Washington & Lee University to put Captain Matt Thompson through a mock interview. Captain Thompson was an Army ROTC instructor and 2002 graduate of the Institute. I had never met him before and was pleased to have the privilege to assist him in this application process phase. After learning of his impressive credentials, I ask him why he had not chosen to go to West Point for free. His response told me a lot about his character and what he valued. Matt told me that in his high school junior year he learned his mother, Susan Thompson, had terminal ovarian cancer. Thompson realized he and his brother, Joshua Thompson(2004), were her support system (single parent family). Consequently, he turned down his acceptance to West Point in order to attend VMI and to be closer to his mother in Virginia. He never regretted that decision. It was his mother's limitless love and selfless example that inspired Thompson to always devote himself completely and to "never surrender the faith." The brothers have established a scholarship in their mother's memory. It is used to support families without sufficient funds to attend key VMI events in which their son or daughter is participating.

Captain Thompson achieved at the highest level during these stressful times. He earned the rank of Eagle Scout and while at VMI he served as the Regimental Commander (First Captain and senior ranking cadet) of the Corps of Cadets. As a commissioned Army officer, he attended Ranger School and graduated as the Top Officer Honor Graduate. From there, he served as a platoon leader in the 101st Airborne Division. This experience took him to Iraq, where he served as ground commander of the largest terrorist raid of 2004, capturing over 89 terrorists. For his valorous actions in combat (2003-2004), this extremis leader received the first of two Bronze Stars. In his evaluation report, his commander stated the following about Captain Thompson's performance: "In all cases, [demonstrated] his tactics, stern and confident leadership, and quick decisions." He went on to credit Thompson for saving the lives of soldiers as well as ensuring the mission was accomplished.

Later, Captain Thompson led an Airborne Ranger platoon that served in Afghanistan in 2004-2005. He then returned to Iraq in 2005-2006 to serve as the executive officer of the Ranger Company. In total, Captain Thompson served combat tours of 18 months in Iraq and 10 months in Afghanistan. While serving in Lexington as an ROTC instructor, he was active in Rotary, founded the Character Counts! educational program for school children, and served as an assistant Scoutmaster in the Boy Scouts. In 2008, Captain Thompson was admitted to the Harvard Business School as an MBA candidate.

Neff and Citrin point out that people are "inherently unpredictable, and uniquely individual," and thus there can be no guarantee that the same ten traits will insure success for every student of leadership. As a matter of fact, you might have noticed that integrity did not make their top 10 list. This may explain why Dennis Kozlowski (Tyco) and Ken Lay (Enron) made their top 50 business leaders list. Both were later indicted on various federal crimes of fraud and other charges.

For purposes of this book, the studies by West and Husted and that of prominent researchers and authors such as Kounzes, Posner, Neff, Citrin, George, and Peters are used to select key characteristics and traits viewed as essential to the success of leaders (business, military, government, and non-profits). With each of these traits, I will describe what kinds of behaviors and practices organizations should expect from their employees when these leadership traits are developed or strengthened through leader development programs such as the one at VMI. Some of the above traits are linked together because of the close relationship one has to the other. This chapter emphasizes these characteristics and describes the importance of each through stories and numerous examples.

Integrity and Solid Values

In the early 1990s, the "big six" accounting firm Arthur Anderson was concerned about the quality of ethics training that its employees were receiving at America's colleges and universities.

An examination of business school curricula nationwide led company officials to believe that not only was the quality suspect, but it was their belief that very little emphasis was placed on ethics when teaching American business students. Thus, Arthur Anderson set out to bring together business professors from around the country to their Illinois training and education center. Their purpose was to develop an ethics curriculum complete with teaching slides, cases studies, and videos for each major business curriculum area. These teaching materials were distributed through workshops to any faculty member who indicated an interest in receiving them. Personally, I felt the four-million-dollar cost of the project was money well spent, and it made me rethink how I integrated business ethics into my disciplines of marketing and leadership. I still use materials from the program.

There are several ironies to this story. First, after using the materials for a few months, I received a letter from Arthur Anderson suggesting that I rethink the use of an introduction to ethics video, which featured grocery entrepreneur Stu Leonard. In the video, Leonard explained the value of ethics and what it means to his grocery chain and to him personally. He commented he wished "my father could see me today." The week before the Arthur Anderson letter was mailed in 1993, Leonard pleaded guilty to income tax evasion for skimming over $17 million dollars in cash during the 1980s with an elaborate computer program that reduced sales totals on an item by item basis. In a second twist of fate, it was the Arthur Anderson firm that was involved in the Enron scandal. As a result of its unethical practices (shredding evidence, etc.), its clients started dropping their services. The end result was a failed company, which no longer exists. Thus, it should be said that it is not enough to talk the talk, but individuals and organizations must also walk the walk. It is my belief that due to the close relationships developed between fellow cadets and BRs, none wants to let the other down or tarnish the name of the Institute. Thus, the ethical behavior of VMI graduates is normally exemplary and well beyond the average college graduate today.

Competence, Creativity, and Intelligence

As an inexperienced 21-year-old Army Signal Corps officer, I quickly learned the importance of competence on the job. After seven months of training to become an isolated combat signal site officer, I was given my first assignment to the Strategic Communication Command Pacific Headquarters (STRATCOM-PAC HQ) at Schofield Barracks, Hawaii. There I served the general staff as a communications status officer. To put it another way, I was one of three rotating and permanent off-duty-hours officers of the day. Our responsibility was to screen all communications traffic between headquarters and Asia, including Vietnam, Korea, Japan, Australia, Guam, Taiwan, the Philippines, and other localities.

If I read a message that appeared urgent, I was to notify the on-call duty officer or the deputy G-3 immediately. This meant making phone calls at all hours of the night to report damage or shutdown of various types of equipment that I had never seen and knew little or nothing about. It seems I was always in the dog house. The messages that I thought were important were not and vice versa. Often when preparing the morning briefing for the two-star commander, I was asked by the deputy G-3 what various broken parts were or what parts were needed to repair a piece of equipment. Of course, I rarely had a clue, for I was trained for battalion and company-size communications equipment and not strategic assets such as satellites and ground receivers. My self-confidence level was pretty low, and I had little opportunity to learn. All the civilian technicians, who might have the answers, worked the day shift and reported to duty after the early morning briefing. The fact that they or more experienced officers and NCOs might not know the answer either did not save the day. I was the responsible person, who was held accountable for not coming up with the correct answers. While at the time that didn't always seem fair, I have since realized that life isn't always fair and each of us must play the deck of cards we are dealt.

Finally, after completing my two-year Hawaiian tour, someone realized that a person with a marketing background and education was better suited for a logistical job versus a highly technical job, where many of the officers had degrees in electrical engineering or computer science. Thus, I was next assigned to the 1st Log Command, which later became the U.S. Army Support Command Vietnam. To prepare me for the assignment, I was sent to the Pocono's Army Depot in Pennsylvania for a three week course. I also trained with the same team that I would work with in Vietnam. After being sent to the Da Nang Army Depot to manage and expedite signal parts to the field, I was finally in my comfort zone and had the opportunity to lead not only approximately 30 U.S. soldiers, but also over 70 Vietnamese civilian employees. With my self-confidence back, my new commanding officer empowered me to run my own shop with minimal interference.

Captain Stewart Husted, the author (second from left) with staff in 1970, at the Da Nang Army Depot, South Vietnam.

I then had the opportunity to be creative and to demonstrate that a college education and intelligence pay off when solving problems under pressure. I was able to devise new storage and inventory processes, conduct inventory audits and spot checks, create a night express system, consolidate multiple sites for single items, reduce pilferage, and most importantly increase our productivity so that we were able to ship more orders in a shorter time. I still made mistakes, but they were far fewer and less frequent. One mistake still sticks out. In the beginning, I didn't know the maximum tonnage for a flat bed trailer. I instructed a soldier to load a trailer as I then watched it collapse from the heavy cargo weight. You can bet I quickly learned the maximum tonnage for all types of trucks and trailers, and I never disabled another one.

Creativity and competence go hand-in-hand. As a cadet, Sir Moses Ezekiel (1866) of Rome, Italy was known for his creativity. He survived the Battle of New Market to become a world-renowned sculptor. After the battle, he cared for a fellow wounded cadet and roommate, Thomas Garland Jefferson (1867), until he died three days later. His compassion and caring for his fellow soldiers made him an impressive role model for others. After the war and with the encouragement of Robert E. Lee (then president of Washington College), Ezekiel pursued a career in the arts. In Italy during World War I, Ezekiel is credited with organizing the American-Italian Red Cross. Later he became the first American Jew to receive international recognition for his creative works of art. Ezekiel studied sculpture after graduating from VMI at the Royal Academy of Berlin. One of his best known sculptures, Virginia Mourns Her Dead, sits in front of Nichols Hall where six cadets killed at New Market are buried. He established his studio at the Baths of Diocletian and was later knighted by the Italians and Germans. He died in 1917 and was interred at Arlington National Cemetery. His creativity served him well as a leader in the world of art and humanitarian causes.[3]

A person's intelligence also plays a key role in successful leadership when coupled with creativity. VMI's 2009 graduation speaker and today's leader of U.S. troops in the Middle East and Southwest Asia is General David Petraeus, a West Point graduate with a PhD. from Princeton University. He is often portrayed in the news media as the military's "warrior scholar." Petraeus

Sir Moses Ezekiel's Virginia Mourns Her Dead
sits in front of Nichols Hall at the resting place of the
six VMI cadets killed at the Battle of New Market.
Cadet musical group plays as a part of New Market ceremonies.

draws from his dual background "to create a leadership style that is at once of a piece with military tradition, yet at the same time innovative."[4] Petraeus's creative counterinsurgency plan coupled with a 30,000 strong troop surge in 2007-2008 began to turn the tide in both the civil war and in the defeat of insurgents fighting American and Allied troops. The statistics for late 2007 through 2008 were impressive. Under Petraeus's guidance, the Army trains differently and teaches its soldiers that how to think is as important as how to fight. The General admits to becoming an adaptive thinker, or one who is willing to be flexible to meet the current conditions on the battlefield. Indeed business leaders can learn much from Petraeus's leadership, for in volatile financial times, creativity, intelligence and passion for a plan can make the difference between failure and success.

Communications Skills

Leaders, who are well educated, demonstrate their intelligence by writing and speaking well. These skills are increasingly important as any employee moves up the career ladder in his or her chosen field. However, one aspect of communications that is often overlooked and is essential to success at every level is the ability to listen, understand, and follow directions. Probably no skill is more easily tested on a Leadership Reaction Course than the ability to listen and follow directions. Participating economics and business cadets in my leadership course are given a two-hour briefing at VMI and a demonstration of some simple icebreakers to lead during a JROTC exercise. They are also given a one hour tour of the course and specific on-site directions to make the written instructions provided (including illustrations) easier to understand.

On a recent FTX, I immediately noticed that one cadet had taken his charges to the same site as another team. If he had failed to remember (assuming he forgot) instead of not listening, he could have looked at his clip board of written instructions and discovered a drawing and a rotation schedule with the exact locations showing where his team was to be at each rotation. At a practice session two days earlier, VMI cadets were given their clip boards with instructions and maps. While this was easily correctable in an exercise, it could be disastrous in a combat situation or in a business scenario. Certainly, if nothing else, followers would quickly lose confidence in their leader. My job as an instructor preparing these future leaders is make them understand the importance of listening and following directions as given and for them to at least ask questions to clarify information they do not understand.

With oral presentations, our faculty has discovered that cadets are pretty competent in the oral communications skills needed to make an excellent presentation or briefing. Using Power Point and other visual aids available, today's leader is expected to make coherent and concise briefings of research findings, operations orders, or to even make speeches to a variety of groups with diverse backgrounds.

George Marshall developed into an excellent public speaker over time. He became adept at one-on-one personal conversations with presidents and world leaders. He understood that the ability to communicate his vision for victory was necessary, and that he had to persuade others that his plan for success was imperative for the Allies to win World War II. Again in 1947, Marshall testified before Congress for six weeks without notes in a successful attempt to sell Congress on the Marshall Plan. He was confident and trustworthy and spoke with a presence of command. Once again at the LRC, evaluators can immediately notice if a cadet is taking command by his voice (strength and enthusiasm).

George C. Marshall testifies before Congress in 1947.
Source: Marshall Foundation

VMI alumnus George Patton once said that "Officers [leaders] must assert themselves by example and voice."[5] Indeed, no one was better at using oral communication than Patton to motivate and inspire his troops during World War II. On one occasion Patton told his officers, "I can tell a commander by the way he speaks. He does not have to swear as much as I do, but he has to speak so that no one will refuse to follow his order. Certain words make you sound like a staff officer and not a commander. A good commander will never express an opinion! A commander knows! No one cares what your opinion is! Never use the words, 'In my opinion, I believe, I think, or I guess.' Every man who hears you speak must know what you want. You can be wrong, but never be in doubt when you speak! Any doubt or fear in your voice and the troops can tell it. Another thing. Never give a command in a sitting position unless you are on a horse or on top of a tank!"[6]

Written communication is also important. When recruiters come to campus, faculty members invariably ask, "What is the one skill students need to improve?" The answer is almost always the same … writing skills. Whether the college or university is VMI or the University of Virginia, the answer is perhaps influenced by the growth of informal communication such as emailing and text messaging. For whatever reason, students are losing the ability to write concise, clear, and understandable written communications without grammatical, punctuation, and spelling errors, or even how to hand write a legible message.

Colleges and universities are attempting to improve writing skills with programs such as "writing across the curriculum." Having students write more term papers is not the answer, if recruiters are asking for a change in current writing habits. While the efforts to use non-English department faculty to emphasize writing in regular classes have merit, most students are still not trained how to write memos, business letters, short reports, technical reports, and proposals. Students at any college need to know such things as when a business memo format is appropriate and when a business letter should be used.

At VMI, we are trying to address this problem by having our economics and business students write short reports using statistics as evidence to support their argument or position in the statistics course; in marketing they may write a sales letter or answer a letter of complaint; or in management they might have to write an employee evaluation using a memo format. Regardless of the specific writing weakness, employers should seek employees who have had lots of practice writing. The day of sitting in a lecture hall, taking a couple of tests, and writing a single draft term paper are over. Today's students need to be coached through multiple drafts to teach them that writing is like an athletic skill. It must be practiced in order for them to improve and be competitive in the job market.

Vision, Passion, and Positive Attitude

My experience as a leader tells me that few things are more important for a leader than establishing a vision for his or her organization. Successful leaders have a clear idea where they want to take the organization they are leading. It is the leader's responsibility to communicate from the beginning their vision and to gain buy in from stakeholders. At VMI everyone associates vision with General Peay. When Peay came to VMI in 2003, his mission was to plan and implement Vision 2039, which included an aggressive building program and fundraising effort. Instead of seeing the elephant somewhere way out there in 2020 or beyond, he envisioned an immediate building program with most construction completed within a 10-year period. In order to step up the construction, Peay needed to demonstrate a passion for the vision and be able to articulate to potential donors and other potential funding sources that VMI needed the new and enhanced facilities now and not later. By the end of 2008, much of the new construction and renovation is complete and the Post transformed into a modern laboratory of learning and living.

Another alumnus with tremendous vision is P. Wesley Foster, Jr. (1956). While at VMI, Foster played guard for three years on the varsity football team, served as editor of *The Bomb* (cadet yearbook), and served as a member of the Honor Court. He also was President of the Officer of the Guard Association. After graduating, Foster served as an Army artillery officer for three years. In 1963, he began a career in real estate as sales manager for new homes with the Minchew Corporation. Between 1966 and 1968 Foster served as sales manager for Nelson Realty in Fairfax, and he personally sold over $1 million in real estate each year over the next six years. In 1968, he and a partner founded Long & Foster Real Estate. In July of 1979 Foster became the sole owner, when he bought out his partner. Today, he is the chairman and CEO of Long & Foster Real Estate Companies (real estate, mortgages, insurance), the largest privately owned real estate company in America and the largest real estate company in the Mid-Atlantic region.

Wes Foster is also a true servant leader, who is generous with his time and money. He has served VMI as a member of the VMI Foundation, The Board of Governors of the Keydet Club, and as a member of the Board of Visitors. He is the recipient of the Foundation's highest honor, the Distinguished Service Award. In 2006 the VMI football stadium was dedicated in his honor as the P. Wesley Foster, Jr. Stadium.

When implementing a vision as large as Foster's, it is crucial that leaders are passionate about their vision (rooted in their values) and daily work. This passion translates into a love for the job and consequently yields the energy, drive, and enthusiasm necessary to conquer long hours and new challenges. Students going to college should be encouraged to seek employment in fields they enjoy and even "love." Furthermore, they should exhibit a passion for excellence. When graduates return to VMI, they often come as recruiters seeking likeminded enthusiastic and passionate students, who are seeking a job that will be fun. These sorts of jobs can slow down or even eliminate potential job burnout.

Another visionary with a passion for business is G. Gilmer Minor III (1963). Minor is chairman of Owens & Minor, a Richmond-based Fortune 500 company, which sells medical and surgical supplies. When Minor took over reins of the company from his father in 1981, the firm had revenues of $205 million. By the end of 2007, the company had revenues in excess of $4 billion.

Information Week, twice since 2001, has named Owens & Minor the number one company of technology users. Also in 2001, Minor was named Virginia's Outstanding Industrialist, and in 2004 he was named by Ernst & Young for his "outstanding leadership" with a special Entrepreneur of the Year Lifetime Achievement Award. Besides serving on numerous corporate and non-profit boards, Minor serves on the VMI Board of Visitors and recently stepped down as chairman.

Finally, there is the strength of possessing a positive attitude. Some VMI cadets are notorious whiners. They complain about everything and everybody, and they love to play the victim. Somehow out of this display of gloom and doom, comes another cadet to class who in a matter of minutes has everyone laughing. He or she views the world in a positive manner and everything is great (at least most of the time). Others master the art of sarcasm and can use it in the most comic of ways. Classes with these individuals are a delight, whereas the naysayers can turn around a class and have them spinning everything in a negative light. Those with a "can do" attitude also set the tone when an assignment is given. Their outlook is refreshing and they know how to make the complicated look simple. With answers to a few questions, they are ready to forge ahead and produce the product needed.

Dependability

Teamwork creates a need and an understanding for the importance of dependability. Rats quickly learn through the Ratline and the Rat Challenge that if they are to survive four years at VMI, they must depend on their roommates, dykes, and faculty and staff. When someone tells you they will do something, you depend on them to do it. They assume a personal accountability for accepting that responsibility. Few things can rock the boat more in business than when a colleague doesn't deliver a product (project report, presentation, etc) when promised. If it is a team project, everyone suffers because of the slacker. When I first came to VMI, I noticed no one tattled on team members. I said something about that in class one day, and a cadet answered that VMI cadets have their own way of dealing with teammates who fail to pull their weight. While I am not sure exactly what that comment meant, I will say that when it comes to issues of dependability, I personally hear little whining. In addition, the research of Husted and West, which used successful VMI business executives as a sample, found that dependability was ranked unusually high in importance compared to other executives (non-VMI alumni) from similar studies.

In the Civil War, General Robert E. Lee (a Union Colonel prior to the war) depended heavily on VMI leaders. Early in the conflict, Governor Henry A. Wise sent Major Thomas Jackson to Harper's Ferry along with 22 cadets to pull guard duty and to prevent rioting at the trial and execution of John Brown.[7] At Gettysburg, 13 of 15 of General Picket's regiments were commanded by VMI graduates. At the Battle of Chancellorsville, "Stonewall" Jackson surrounded himself with a staff composed of four VMI faculty (Rhodes, Munsford, Colston, and Crutchfield). At the Battle of New Market, Major General John Breckinridge openly prayed for God's forgiveness as he "called for the boys" from VMI who were guarding a baggage train. The 241 cadets from VMI were able to turn the tide of the battle. The cadets accounted for 27 percent of Confederate losses. In the battle for Lynchburg, Brigadier General John McCausland (1857), an assistant math professor, was sent with a small cavalry brigade in June 1864. He and the cadets were able to stop the Union advance, and the unit then headed north to advance on Washington D.C. and Chambersburg, Pennsylvania. From the beginning of the conflict, other cadets were sent to Richmond to train new recruits.[8]

Loyalty, Concern, and Devotion

Most people think of loyalty as a faithfulness or allegiance to others. Employers expect and some demand loyalty from their employees. One former employer of mine told me point blank in a job interview that if I was not loyal, I would lose my job immediately. This individual did not want me sharing out-of-shop stories about him, his performance, or the decisions made. The bottom line with loyalty is: Do individuals always stand behind and support people they work for or someone they work with? Are they the devoted employees, who will go the extra mile when a special effort is needed such as staying late, or will they go home when their eight-hour day is over? Are they concerned about their fellow workers and do they chip in to assist when someone is behind or absent, or do they let them struggle and allow their workload to increase?

In the Rat system at VMI, cadets learn the true meaning of loyalty. Loyalty should never be blind. Cadets do not protect each other just because they are Brother Rats. Loyalty stops when a cadet crosses the line and violates the honor code. In business, loyalty should stop when an employer asks you to do something illegal or unethical. Loyalty to a military commander does not extend to following an illegal order. Loyalty should be extended to those who are doing the right thing for the right reasons. VMI promotes that true loyalty to fellow cadets is a lifetime concern and commitment shared by a network of connected relationships of BRs and other alumni.

A second kind of loyalty also exists. It is not measured by your commitment to others, but to the principles and convictions that you hold. If you don't believe in stealing, do you take office supplies home such as pens and paper? If you know that you will be fired if you do not commit an unethical act, would you resign or would you rationalize the act as necessary to keep your job? Loyalty to one's convictions is perhaps the toughest type of loyalty. In today's society each of us will be constantly tested. Can we say we are truly loyal?

To aid cadets in testing the waters of loyalty, I give them case studies in their leadership classes, which ask them to make decisions and to justify their answers. While there is no way the classroom can address every situation related to loyalty, faculty can expose students to the realities of the real world and make sure they know that gray areas can sometimes become very dark, very quickly.

Persistence and Determination

The Coalition effort in Iraq can only be described in terms of pure persistence to stick with the unpopular job of defeating insurgent forces in an extreme environment of heat, dust and sand storms, and dreaded IEDs. This war, thus far, has lasted more than six years and called on our troops to pull multiple tours and to "soldier on" for as long as 15 months during the surge of January 2007 to March 2008. Our soldiers' determination to achieve victory has been an incredibly difficult mission with pressure coming from Congress, the media and others to bring the troops home without defeating al-Qaeda or the Taliban in Afghanistan. In many military quarters, these terms spell retreat. The Marine Corps has banned the word retreat from its vocabulary and preferably uses terms such as rearward action or withdrawal. These terms imply a more temporary movement off the front lines. Never will you hear a Marine use the word defeat, because a Marine never gives up. A Marine is taught to keep pushing, pushing, and pushing some more until through pure persistence and determination a mission or victory is achieved.[9]

These troops develop a brotherhood of tight loyalty. They fight not just for their nation and their convictions, but they fight for each other. This type of loyalty and determination requires tremendous sacrifice, not only in terms of time and bloodshed by soldiers, but on their families and loved ones. Like many men in World War II, another generation of men and women have married and then rushed off to war, not to see their spouses for long periods. Those with young families can hardly recognize the young baby who was crawling when they left and is now walking and talking. Those families, who can, band together on post in unit support groups. A few members of the unit are left behind to handle communication with families and attempt to provide for their many needs. Often things change back home and military personnel miss funerals, weddings, graduations, and even the births of their children. Sometimes they have to deal with spouses, who are unfaithful or can't deal with the challenges of raising children on their own. It is no wonder that the divorce rate for military personnel is very high.

Selfless Service and Sacrifice

Matthew 23: 11-12 (NKJV) tells us, "He who is greatest among you shall be your servant. And whoever exalts himself will be humbled, and he who humbles himself will be exalted." Jesus both warns and challenges us in this *Bible* verse that as leaders it is our responsibility to serve others and to avoid self-serving actions, which are not done as acts of heart and compassion. Do we serve in a leadership position for the esteem and power of the position, or are we leading as an act of service? This is a question that cadets must ask at least once a year. Do they want to serve the Corps as servants, or do they want to be a "ranker" because it gains them recognition and makes them look and feel important? Cadets must learn that life is a journey, and at some point they must be motivated by what they can give instead of what they can get. Regardless of age, one reaches true adulthood at the point when pride is overcome and self-promotion is defeated for the good of others.

VMI alumnus Carter L. Burgess (1939), former Chairman of the Board of AMF, said in his Founder's Day address in 1966 that "this country should never get so wrapped up in technology and the miracle of missiles to forget the continuing greater requirement for our youth and manpower to speak up and stand for service and duty - and here's where VMI excels."[10]

Selfless service is exercised every day at VMI. Rather than choose the obvious such as Corps leaders and sports captains, I'd like to focus on a true group of selfless servants, who are seldom recognized for their contributions not only to VMI but to the community of Rockbridge County. One cadet organization on Post is completely dedicated to helping others. Our EMTs are constantly sacrificing their time and personal dollars to provide emergency medical care to cadets on Post. Some also serve on the local city and county volunteer squads, which make emergency runs all over Rockbridge County including the busy and very dangerous Interstates 81 and 64.

One cadet, Ryan Corcoran (2008), was having difficulty staying awake in my class, and he was late with an assignment.

When I questioned him, I learned that Ryan was a volunteer EMT and was often called out late at night when he was sleeping or studying. Ryan, Paul Hiner, and Tim Gallinia from my leadership class and other EMTs have volunteered their time to serve the emergency needs of participants at the Goose Creek Outdoor Education Center. Their service to this non-profit has been invaluable.

Another VMI EMT, Sergeant Ryan Doltz (2000) volunteered for his home EMT squad in New Jersey after graduating from VMI. While a student at the Institute, Doltz was a member of the band and Army National Guard Battery A of field artillery from Martinsville. In 2004, he deployed as an MP with the 112th New Jersey Army National Guard Field Artillery. On June 5, Sergeant Doltz was killed by an IED while on patrol. Some may remember Doltz as one of four cadets featured in a Norelco shaver ad campaign for the company's Advantage wet-shave electric razor. In the ad, a gruff drill instructor pinched his cheek and called him "baby face." Doltz was known by his friends as being "an all-around good guy, who would do anything for you."[11]

Captain Lowell T. Miller II (1993), a Michigan National Guard member, made the ultimate sacrifice for his country. Miller was killed in action on August 31, 2005, by small arms fire. He had volunteered for duty in Iraq and was attached to the Mississippi National Guard during its Iraq deployment. He was on a mission with an Iraqi army unit, which he was training. Miller left behind a new bride and two step-children, his parents, and two siblings who also serve in the military. In a letter to his father, he said, "Dad, I serve so others don't have to. You taught me to be a leader, to stand up and sacrifice so others would not have to; yet we do. You taught us well."[12]

Andrew Dernovsek (2006) served mankind in a different venue. Andrew traveled to the mountainous terrain of Lesotho, Africa as a member of the Peace Corps. His background as an international studies major and his training in the adverse conditions of VMI were most useful to him in his many duties and responsibilities. Dernovsek claims on his web blog that

VMI helped him develop "nerves of steel." He needed them as he rode on horseback over rough terrain to visit isolated villages, where HIV/AIDS education was badly needed. He also assisted villagers in gardening, pig breeding, and water projects, and taught English and occasionally math and science to boot. Dernovsek stated in an interview for the VMI news service that "I serve as an ambassador for the American people ... I am the only American that the vast majority of people in my area will ever meet. I serve to show them what an American really is."[13]

In 2007, another graduate demonstrated his will to serve others in a different but important venue. Rich Meredith (2007), the top ranked student in Economics and Business with a 3.9 GPA and a third-generation cadet, surprised the faculty when he informed us that instead of rushing to graduate school or taking a high paying job in the investment world, he would pay back his debt to society by volunteering for the low paying but prestigious non-profit Teach for America. While a cadet, Rich was no slouch in the service area. He served as vice-president of his class, a member of the General Committee, Regimental S-2 captain for academics, Group B Investment Club president, and G Company Rat Challenge lieutenant.

Teach for America is dedicated to providing young college graduates from many of America's best colleges to teach in socio-economically challenged regions of the country. Today there are 5,000 teachers serving two year commitments in over 1,000 schools in 26 regions of the nation. It is hard to imagine a better candidate for such a position. Teach for America's core values are a great match for those held by Rich and VMI.

Meredith has been in regular contact with faculty and staff at VMI and has shared his triumphs and challenges in rural Mississippi. He has described conditions which at best are bad. The interest in academics is low and the atmosphere is one of high tension. In an email sent to the faculty he shared, "My biggest challenge is classroom management and just the cultural difference between students and myself...In all my classes I have to deal with talking, disrespect, profanity, throwing things, etc. on a daily basis. But I'm starting to look past some of that to

Cadet Rich Meredith (center) and Brother Rats (2007)
Ryan Smith (left) and Maury Denton (right).

some really great students. I hope to build positive, influential relationships with my students – the troublemakers too. Above all else, I think teaching here will show me how to love and care for people in a different way."[14] Needless to say with his positive attitude, Meredith will be successful in Mississippi or any other location where life takes Rich.

The schools in his area have few resources such as paper, pencils, and other supplies. To assist Rich, several faculty and staff collected money to purchase supplies and other needed items to assist him. Donna Potter, departmental secretary, working with members of her Colliertown Presbyterian Church, purchased supplies and sent them to Meredith. During the 2007 Thanksgiving break, Rich and his parents came to Lexington and held a reception for the faculty and staff, which assisted Rich in his pursuit of educational equality, excellence and selfless service.

Another graduate, Frank Louthan (1966), used his extensive business experience as a pharmaceutical CEO and president to found a non-profit to teach business principles in developing nations. His organization, HighMark, strives to mobilize thousands of American business persons with their skills, experiences, and life values and to instruct others in the principles of business leadership. Their goal is to transform lives and nations through the impact of successful businesses, built on the business principles taught in HighMark seminars. Volunteer instructors are currently teaching in Kazakhstan, Azerbaijan, Russia, Ukraine, Philippines, Vietnam, Greece, Cameroon, Ecuador, and Macedonia.

George Marshall was also known to be a servant leader. Marshall was a very selfless man, who always put others before himself. He was willing to sacrifice to win the war and to serve his nation. One classic and well-known example involved the selection of a D-Day commander. Marshall wanted the position badly and was preparing for the assignment. Many in Washington argued that Marshall should remain as Army chief of staff, because no one had the global perspective and experience that Marshall possessed. When asked which position he wanted, Marshall told Roosevelt that he wanted to serve where he was needed most. That was enough for Roosevelt to name General Dwight Eisenhower as commander of European troops and eventual D-Day commander. Had Marshall told Roosevelt that he wanted the D-Day command, the world might have become a different place. There might have been no Marshall Plan that saved Europe, and Marshall not Eisenhower might have become president of the United States. After the war, although sick and worn out, Marshall served his country again in a variety of high-level civilian and cabinet positions.

When I think of other sacrifices made by VMI cadets and alumni, my mind immediately recalls the cadets at New Market, but I also think of a First Class cadet, Nick Wegener (2008), who was a student in my leadership course in 2007 and a senior member of the football team. In the beginning of the semester, Nick would "unofficially" ride the elevator (off-limits to cadets) up with me in the morning on the way to class. He explained to me he was out for the season, because of serious knee injuries requiring surgery and that he had developed arthritis so bad that he was in constant pain. At this point Coach Jim Reed was using Nick to film games.

Later in the season, Nick decided the team needed him on the field dressed out and practicing. With determined passion showing in his eyes, Nick shared with me that he felt his younger teammates needed an on-the-field role model to demonstrate that they too needed to sacrifice if VMI was ever going to win the important football games remaining in the season. Of course everyone was concerned about his decision, because his injuries were serious enough that Coach Reed would never play him again in a game. Some players and cadets could not see Nick's point and ridiculed him for being "stupid." Nick, however, never gave up and continued practicing to the last game against the Citadel. While I also questioned Nick's decision, I never questioned his heart or how much achieving success on the playing field meant to him. Nick was determined to complete what he had begun in his freshman year.

Fairness in Decisions

Life is not always fair. To residents of New Orleans and coastal Mississippi, it may not seem fair that their homes were destroyed by Hurricane Katrina or that brush fires in California randomly spared some homes while destroying others around them. To some it may not seem fair that our soldiers and their families are sacrificing and enduring extreme hardships with multiple, dangerous, and long-term combat tours, while at the same time 99.5% of the nation goes on with few cares in the world beyond what DVD they watch tonight, what meal they plan to eat, or what car they drive. Never mind the argument that today's soldiers volunteered. Given that the U.S. is at war, it seems very unfair to many that so few are burdened with such an awesome responsibility that ultimately affects all of us for generations.

At VMI, a few cadets and graduates may feel it is unfair that approximately 50 percent of the Institute's graduates commission in an armed service, while their classmates with the same degrees and diplomas rush off to Wall Street or some plum civilian or government job. While this is may be a hornet's nest topic, one cannot help notice the sacrifice and commitment made by many of our cadets and how it separates them from their classmates in untold ways.

What I learned from my cadets was that they do understand that life is not always fair. It isn't fair if a classmate gets cancer or their mother dies or a girlfriend dumps them right before Ring Figure Weekend; but that is a different type of fairness. What cadets expect and demand is fairness in the decisions that affect their daily lives as cadets. As a faculty member, I had to be fair in my grading system, my tests and exams, and how I graded essays or other papers. What a faculty member soon learns is that despite strict privacy laws to protect students, many cadets share every grade and every professor comment provided, orally or in writing. Thus, if I took five points off on an essay question for something Cadet Smith got wrong, I'd better take five points off for Cadet Jones as well if the same question was missed.

Sometimes after hours or days of subjective grading, it is difficult to remember what you may have taken off earlier on another student's test for a same or similar incorrect essay answer, especially when you are trying to give them partial credit for what they did get right in a generally wrong answer. One thing for sure, they know and will think the professor is being unfair if the points deducted are not equal. Most cadets will quickly bring this to the attention of the professor and most will remain respectful if they don't get the response they want. Many cadets who challenge points deducted are often uninformed, and quickly after an explanation understand why they are wrong. On the other hand, if I made a mistake, I must be willing to change the grade on the paper or exam.

The same applies to the barracks. If an Officer in Charge (OIC) is inspecting rooms, nothing would create a protest more if demerits awarded for an offense were not the same for every cadet not meeting the room standard. Thus, if Colonel West found a dirty room sink, he should not give five demerits to Cadet Dufus, three to Cadet Pet, and six to Cadet D'Nothing. If Cadet Rambo was caught without a shave in morning formation, it would be unfair if his platoon leader skipped Cadet GoForre because he was a close friend. Cadets expect fairness in the decisions that affect them on a personal level, but understand that everything in life is not a human decision and our lives takes different paths.

In the business world and in the military, the issue of fairness is also important to employees. No one likes it if the boss's son or daughter gets promoted with no or little experience, or if a male is promoted over a female with equal qualifications, or if an older person is passed over for younger blood. In personnel issues, fairness often is marked by acts of discrimination for reasons of age, gender, race, nationality, or religion. Discrimination in the workplace has no place. Fairness at all levels and all issues is expected and should be demanded.

Courage and Self-Discipline

Courage requires self-discipline. The disciplining of oneself is an achievement that is not easily learned. It is the result of self-sacrifice and determination to make things right. Every cadet who attends VMI learns the importance of self-discipline; however, personal courage is harder to demonstrate. While courage may be proven at VMI through endless moral decisions, not all cadets have the opportunity to publicly demonstrate they possess it. Some prove their courage on the battlefield (see Table 4.1) and others through extraordinary acts, often unrecognized.

Table 4-1: VMI Medal of Honor Recipients

Clarence Edwin Sutton 1890	USMC	Boxer Rebellion: China
Charles Evans Kilbourne 1894	USA	Philippines
Cary Devall Langhorne 1894	USN	Mexico
William Peterkin Upshur 1902	USMC	Haiti
Richard Evelyn Byrd Jr. 1908,	USN	Polar explorer and aviator
Edward R. Schowalter Jr. 1951	USA	Korea
Adolphus Staton 1900	USN	Mexico

One alumnus, who could not tolerate unfairness or discrimination, was Jonathan M. Daniels (1961). In his young life (26 years), he was able to muster a great deal of courage to participate in the southern Civil Rights Movement. By courageously serving African-Americans, he ultimately sacrificed his life for them. Daniels from Keene, New Hampshire was valedictorian of his VMI class and a man of great self-discipline and personal courage. Although Daniels did not like VMI, his criticism was often humorous. Like George Marshall, he found the Southern culture very different from New England. As a Rat, he had a difficult time adjusting, but eventually adjusted and won the respect of his classmates. As time evolved he became a skeptic and questioned the value of everything from military life to his Christianity. Viewed by his classmates as a true intellectual with a great curiosity, Daniels was known to be very compassionate and a true "Rat daddy".[15]

Senior picture of Jonathon M. Daniels (1961)
who went on to become a martyred civil rights leader. Daniels was killed
on August 20, 1965
in Haynesville, Alabama.

After graduating from VMI, Daniels entered Harvard to study English Literature on a Danforth Fellowship. Although Daniels questioned his religious faith while at VMI, he felt his faith restored while attending a service at the Church of the Advent in Boston. His conviction was so strong that he felt a calling to serve God, and soon afterward he enrolled at the Episcopal Theological School in Cambridge, Massachusetts in 1963.

In March of 1965, Daniels went to Selma, Alabama at the request of the Rev. Martin Luther King with other students and clergy to take part in a march from Selma to the state capital in Montgomery. Realizing that they were needed for longer than a weekend, Daniels and a friend, Judith Upham, returned two weeks later. The two received permission from their school to study on their own and to return at the end of the semester to take exams. Daniels stayed with an African-American family and devoted his time to integrating the local Episcopal Church with groups of young African-Americans. After his exams, he returned in July to continue his work and to tutor children, register voters, and assist poor locals in applying for federal and state aid.

On August 13, 1965, Daniels and 29 other protestors were arrested for picketing white-only stores in Fort Union, Alabama. With the exception of five juveniles, the group was held for six days in a tiny jail cell in Haynesville. On August 20, the group was released, but had no transportation back to Fort Union. Daniels and three others then went for cold sodas at a nearby store that served nonwhites. They were met at the door by an unpaid special deputy sheriff, who told them to leave or be shot. After a verbal confrontation, he leveled his shot gun at seventeen year-old Ruby Sales. Daniels pushed her aside and was immediately hit by the blast from the shotgun. He died instantly. The others ran and a Catholic priest, Richard Morrisroe, was shot in the lower back. Typical of the era, the killer was found not guilty by a jury of 12 white men. The defendant claimed it was self-defense and that Daniels had a knife which was never found.[16]

It took the death of Daniels to awaken the Episcopal Church to the injustices of segregation and to the civil rights movement. In 1991, Jonathan Daniels was named one of 15 modern-day martyrs. August 14 was designated as a day of remembrance for the sacrifice of Daniels and other civil rights martyrs. In turn, VMI dedicated one of four named arches in the Barracks in his honor and established the Jonathan M. Daniels '61 Humanitarian Award. The first award (2001) was presented to former President Jimmy Carter and the second to Civil Rights leader and Ambassador Andrew Young (2006). Daniels has been the subject of two books and a documentary.

Major Paul Syverson (1993) proved his courage on the battlefields of Afghanistan and Iraq. Syverson, a member of the 5th Special Forces, was one of first Americans to be seen on TV in the battle for Afghanistan. His unit responded to the prison uprising at Qalai-Jangi. The Special Forces team captured American John Walker Lindh, who was fighting with the Taliban, and recovered the body of CIA agent Mike Spann, who had been killed early in the uprising. Major Syverson was killed by an insurgent mortar round on June 6, 2004 during his third tour in Iraq. In an AP article, Syverson was recognized for his well-known leadership skills. Lieutenant Colonel Christopher Haas said, "His soldiers would stop what they were doing just to speak with him or be near him if only for a few moments."[17]

Another alumnus (U.S. Naval Academy graduate) Richard E. Byrd (1908) was a pioneering American polar explorer and aviator. Byrd was another extremis leader who took great risks during his life's quest for adventure. On May 9, 1926, he claimed (disputed) to have flown over the North Pole and was subsequently awarded the Medal of Honor. In 1928-1930, Byrd led the first American Antarctic expedition and constructed the base camp "Little America." After the first winter, Byrd along with pilot Bernt Balchen and co-pilot Harold June, flew to the South Pole and back in 18 hours and 41 minutes. During the flight, the aircraft developed an oil leak. In order to gain enough altitude to circle the Polar Plateau (Queen Maud Mountain Range), the three had to jettison their emergency gear and empty fuel tanks. For his flights to the South Pole, he was awarded the Navy Cross.

Byrd returned to the Antarctic four more times. In his second expedition, 1933-35, he spent five winter months alone operating a meteorological station 120 miles from the base station. After receiving unusual radio transmissions from him, a doctor and two expedition members made three attempts to reach Byrd. Finally, they reached him, but he was in poor health because of carbon monoxide poisoning from a poorly ventilated stove. Rear Admiral Byrd's courageous deeds on behalf of the United States opened a new door for discovery, science and technology. [18]

Conclusion

Being an authentic leader requires more of us than the average person can ever contemplate. Over 200 traits have been identified that successful leaders possess; however, in the end, only a few really make a true difference. This chapter has summarized the literature and research to offer an examination of the leadership qualities one should expect from a 21st Century leader. At VMI, a model exists and an effort is made to provide cadets with a series of leveled leadership experiences and classroom instruction to assist in developing the type of leaders needed for the future of our nation. While every cadet will not measure up to the standards described in this chapter, every cadet has an opportunity to achieve those standards and develop those leadership qualities. If not at VMI, perhaps they will later emerge to be the leader within that developed over their four or five years at the Institute. Few colleges and universities can make that claim.

5

INTEGRITY FIRST & ALWAYS
MORAL & ETHICAL DEVELOPMENT

"The tradition and standards ... have permanently endowed the Institute with a legacy for the development of future citizens having that stamp of character necessary to the maintenance of genuine democracy."

George C. Marshall,
Army general and statesman

VMI is a place where stories of lore are often and widely told by alumni to anyone who will listen. Many family, friends, and others not associated with the Institute frequently dismiss these stories as myth or war stories told to embellish the VMI experience. While that may certainly be true in some cases, I have personally witnessed several incidences from 2002-2009, which strengthen my opinion that there is always some truth at the heart of corps lore. For instance, I'll never forget that in my first year at VMI, a colleague from another college came to visit me. As I was giving him a tour of Scott Shipp Hall, he noticed an expensive calculator sitting on a desk in a classroom. Keith told me I should take the calculator down to the office so the secretary could secure the item. He then picked up the calculator and handed it to me. I took it and immediately placed it on a media cart and reminded him that he was at VMI. I simply did not need to take it to the office, which was probably the last place a cadet would look for something misplaced.

Approximately, eight months later, I was conducting a spring open house for prospective students and their parents. I was telling this story, when a parent pointed to the media cart. There sat the calculator, just as I had left it. Two years later when teaching in the same room, it was still sitting there. On two other occasions, I have witnessed a $20 bill and a $5 bill, thumb tacked to the departmental bulletin board in our hallway. Others might notice cadet covers (hats) that are piled on dining hall tables with IDs, cash, keys and other valuables visible inside the covers for all to take. There are no finders keepers at VMI.

The Importance of Character

If these stories strike you as unusual, they should. While they are not unusual at the Institute, few colleges would find students pinning up dollars found on the ground, if for no other reason than no one can prove they lost it, so why not keep it? If I were to take the $20 bill off a VMI bulletin board, it would be assumed that it was mine. There would be no questions asked. I would be trusted as a person of character not to steal. My behavior would be consistent with my character. *Leadership should be character in action.*

Twentieth century English author and critic, Lord G. K. Chesterfield, once commented to his son that character "is the sure and solid foundation upon which you must both stand and rise. A man's moral character is a more delicate thing than a woman's reputation for chastity. A slip or two may possibly be forgiven her … but a man's moral character is irreparably destroyed."[1]

Individuals of character must seek out the truth and act on their beliefs. In the end, it takes leaders of character to create and lead organizations of character.[2] At VMI everything is about honor and trust, both characteristics of what defines a cadet's individual character. Character is our consistent ethical behavior that others can expect from us. According to University of California leadership guru Warren Bennis, "Character is the key to leadership." Bennis draws his conclusions from a Harvard study that indicates that 85 percent of a leader's performance depends on personal character.[3]

VMI does a great job of recruiting students who have been positively shaped by family values. That fact should never be downplayed; however, anecdotal evidence abounds that the VMI experience also strongly contributes to the character development of many cadets. In my 2002 leadership survey, many alumni offered responses to the question: What role did VMI play in the development of your character? Travis Russell (1955) responded that, "I did not fully appreciate my VMI experience until way into my military, business, and family life. I have always been proud of graduating from VMI but little did I realize how much values and training helped shape my character and leadership abilities."[4] Carlos Zamora (1984) believes that "VMI was the most transformational and defining experience in my life."[5]

Just as many a son or daughter doesn't want to let their parents down when making key life decisions, so are there similar feelings about VMI. Jim Joustra (1976) believes that "Coupled with the values my parents gave me, VMI set a standard for honor, service and excellence by which I have lived my life. I have always felt that my individual performance reflects on VMI and I cannot let the Institute down." Joustra goes on to say, "Several years ago I was faced with a decision to either lose my job or follow an order/directive to break the law. I chose to lose my job rather than sacrifice my integrity and let down my VMI Brothers. I then had to move my family over 1,000 miles to my new job. It was the right decision and I have no regrets. My career has gone well."[6] These comments represent the scores of responses received. Only three respondents felt that VMI did not play a direct role in their character development.

Despite the Harvard findings, Americans exhibit contradictory beliefs and consequently contradictory behaviors when it comes to character. While we desperately seek and demand leaders of character, we often select leaders of questionable character. A question often asked is, "Does it matter?" President George W. Bush ran on a platform of values, but in the end his presidency will likely be defined as less effective than Bill Clinton's term in the White House. Dr. James Hunter, a professor at the University of Virginia's Institute for Advanced Studies in Culture and author of *The Death of Character* and *Culture Wars*, says that "It is apparent that we as a culture do not really want it [moral leadership]."[7]

Leaders trying to distinguish between right and wrong or good and evil on moral issues are often branded as intolerant. Hunter's research on character in leadership found that 54 percent of those surveyed agree that a politician can be effective, even if he has little personal character. Furthermore, Hunter claims in an interview with Les Csorba (partner with Heidrick & Struggles executive search firm and former White House advisor) that Americans are "petrified of such moral leadership because of the very demands it would place on each of us."[8] At VMI moral leadership is expected from day one at the Institute. It is an expectation now and forever. Anything else is not tolerated.

So again I ask the question, "Does personal character matter?" Can we expect a new era where personal integrity and character reign? Hunter is not optimistic. His research indicates that Generation X is more likely to put happiness above integrity, to "live for tomorrow," and live by the dictum "eat, drink and be merry." Other studies indicate that today's Millennium Generation is even more into themselves and likely to engage in cheating and lying than past generations. Their approach to leadership decisions is also more likely to be pragmatic than based on morals.

As a personal observation, business schools are working hard to integrate ethics into the curriculum. One such way is through case studies, some real and some contrived, which pose ethical dilemmas for those in leadership roles. Often I have been amazed at the responses that students at both the undergraduate and MBA levels (non-VMI students) have offered as solutions to what I believed were straightforward right and wrong issues.

In 2007, I authored a business ethics case published in the *Business Case Journal* about the management and fundraising for construction of the National D-Day Memorial in Bedford Virginia.[9] A dream by a local D-Day veteran to build a statue to honor those who lost their lives on D-Day developed into a $25 million magnificent project that took over a decade to complete. The memorial was dedicated on June 6, 2001 by President Bush. The dedication was attended by over 20,000 visitors and shown live on CNN.

Claiming health problems, the foundation president resigned several weeks later. Unknown but to a few, he had withheld from his board that project overruns had cost the organization an additional $7 million. In June of 2002, he was accused by the Federal government of lying to his board, donors, and lenders and fraudulently using some designated gifts to complete the memorial. For example, the foundation president had used a $100,000 gift by Stephen Spielberg to build an education center theater to pay off general debts; and he used a $2 million non-binding pledge, which failed to materialize, as security for a $1.2 million bank loan. So why did the foundation president make these decisions? His response in court and to media was that he owed it to the D-Day veterans to complete the memorial on time. Over 1,000 World War II vets die each day, and there was a real sense of urgency to complete the project for the scheduled dedication by Bush.

Let's do a reality check. Where do you stand on moral versus pragmatic leadership? Would you find the foundation president guilty of fraud and other charges, or would you vote to find him not guilty because his alleged lies and acts were not done for personal gain? You might find it interesting that all charges were dropped in 2004 after two hung juries from two different cities in Virginia. The majority of jury members in both trials voted for acquittal. Welcome to pragmatic leadership. To make it more interesting, the U.S. Justice Department recently investigated the Federal district attorney for misconduct in the trial. He was accused of intimidating a witness and withholding exculpatory evidence; however, the investigation revealed no evidence to substantiate the charges.

So what are students to conclude? Is it okay to lie because more people benefited than were hurt, or is it always unacceptable to lie because people must trust you and what you say is a statement of your character? As a professor and father, I must be careful when teaching ethics or raising my children to teach them that although different philosophical approaches to ethical decision making may yield different responses, there are actions, which are more acceptable and morally correct. While that may sound like a no brainer, you might be surprised by what is sometimes taught in classrooms across the nation under the guise of "academic freedom." For instance, Ward Churchill of the

University of Colorado and a few other academics claim the U.S. government conspired to bring down the Twin Towers on 9/11.

Why do these ideas hatch? A likely reason is because our government and corporate leaders are not trusted. Once again trust is an integral part of character. If leaders are not demonstrating moral leadership, then people will not trust them. Unfortunately, the media regularly informs us that our government and corporate officials lie, cheat, and steal when it benefits them.

VMI cadets are taught from day one that VMI cadets will not lie, cheat, steal, nor tolerate those who do. On the first night for new "matriculates," you can hear hundreds of them reading the honor code aloud and committing it to memory. Their honor code demands nothing less and over time and after graduation their personal honor code (as opposed to one followed because of punishment or dismissal) will kick in; and hopefully they will not want to be associated with those who cheat, lie, or steal.

Marine Captain Brian C. Collins (1994), while deployed in northern Kuwait, was quoted by the *Richmond Times-Dispatch* in a story by an imbedded reporter, Rex Bowman, that his VMI education afforded him the ability to "learn the value of honor and integrity, to learn that hard work is rewarded – that's invaluable." Collins' battalion commander said that Captain Collins was "very, very, very technically proficient. I have great trust and confidence in him."[10]

In 2001, I served as a group facilitator at the West Point (USMA) National Conference on Ethics. I was surprised that my group of 14 students felt it was equally important that faculty and staff be held accountable for their words and actions in the classroom. In other words, they wanted an honor code for faculty and staff at their colleges. At VMI, the Board of Visitors in 2007 adopted a Code of Ethics for all employees of the Institute. The code contains a statement of ethical values and standards for ethical conduct.

What can the public expect from young men and women who are trained to be honorable leaders of character? The answer is found in the way they handle tough moral and ethical issues. Jim Hackett, CEO of Steelcase, the office furniture manufacturer, followed the advice of his mentor, hotelier Bill Marriott, who told him to build his company on "unyielding integrity." Ten years later, Hackett was faced with a dilemma, which eventually cost his company millions. Hackett discovered that Steelcase had a potential problem with a new product line of office panels that could also be used for floor to wall dividers, plus for their original use as waist high dividers for office cubicles. When the company discovered that the fire retardant regulations were stricter for walls, some in the company on impulse wanted to ignore it. Even some customers said it didn't matter to them and encouraged the company to continue selling the successful product line. Because fire codes vary in different localities, this problem fell into a big gray area. In the end, Hackett decided if Steelcase was going to be run on "unyielding integrity," he must replace the panels with units that met the stricter codes. That decision cost Steelcase $40 million. Hackett and his executives lost their annual performance bonuses that year.[11]

Another example is offered by William Diel, Jr. (1952), "During my first year of working in industry, a contractor had not performed an important function according to specifications. I was asked to ignore the infraction, which I declined. Later in my career, I realized that my action had established a reputation for honesty and integrity that remained throughout my career."[12] Another grad, Gerald Quirk (1962) spoke of disagreeing with a key executive on a key corporate decision. The company owner tried to fire him, but held back when the employees demanded his retention. Quirk was right in his assessment and his plant showed a high profit, while three years later a sister plant following corporate guidance failed. He said, "Doing what I think is right regardless of the risk was directly linked to my VMI experience."[13]

VMI goes to great lengths to protect its image of producing leaders of character. The symbol of the VMI experience is the VMI class ring, around since 1848. It is a constant reminder to every alumnus of what a VMI education signifies. Each class ring represents a "coded history" of symbols representing the experiences of each class. For example the class of 1998 included a small scroll to represent the Institute's long legal battle to prevent the Institute from going coed.

Many a cadet has married their VMI Ring Figure date.
Source: VMI Communications & Marketing

Ring worship is common at service academies and military colleges. West Pointers, also known for their rings, are commonly called "ring knockers." There is almost a competition between classmates to see who can purchase the biggest, gaudiest, and most expensive ring in the class, and which military college can produce the largest ring each year. As a Virginia Tech graduate, I know how important the ring was that I received at the end of my junior year. At VMI a Ring Ceremony culminates in the fall Ring Figure dance and other celebratory activities.

Despite the importance of my ring, I never wear it now. I am not sure exactly why, but I suspect because my fingers have gotten fatter and maybe because so many people have Tech rings now that I no longer feel it puts me in an elite group. Certainly, no one can look at my Tech ring and tell I was in the VT Corps of Cadets. On the other hand, I have never met a VMI graduate who did not proudly sport his or her ring. Cadets know that only a small number of people have earned that privilege. Some graduates continue to wear their rings on their wedding finger after they are married. I once asked a cadet who was constantly complaining about VMI life, why he just didn't leave and go to a school that would make him happier. He gave me the most incredulous look, and said as if I were totally stupid, "To get my ring, sir."

The husband of teacher- astronaut Christa McAuliffe, Steven McAuliffe, is a VMI graduate, class of 1970. Christa McAuliffe took her husband's ring aboard the Challenger space mission, which exploded on January 28, 1986. To ensure that McAuliffe had a ring, his classmates had a replica of his ring made for him. The original ring was later recovered in good condition and the replica is now displayed in the VMI museum. After graduation, Steven McAuliffe served as an Army JAG officer and in 1992 became a Federal judge with the U.S. District court in Concord, New Hampshire. McAuliffe credits his VMI background for helping him through the tough times following his wife's death.

Another graduate and assistant commandant, Major Mitch Fridley (1989), was asked by the class of 2008 to carry a bar of metal into combat, which was then melted down upon his return and put into the rings of each class member. Fridley served as chief of the Force Management Section, J-3, in Baghdad from 2005-2006. For his "meritorious conduct and outstanding service," he was awarded the Bronze Star.

In 2007, a story ran in the media about a 40-year-old high school track coach who married a 16-year old member of his track team. A photo appeared of the man in news stories worldwide. He was sporting a VMI basketball t-shirt. Immediately the phone calls and emails from alumni flooded VMI demanding to know what connection this character had to do with VMI. Was he an alumnus? Family member? Fortunately, the answer to both was negative. There was no connection; however, the VMI community was horrified that one of their own could have possibly committed an act of bad character.

Assuming that a pragmatic world today influences our leaders and our culture, the question becomes how do we find and produce leaders of character who put the greater good of society before their personal dreams of monetary gains and the good life? I believe this is where VMI comes into play. For over 170 years, VMI has worked to generate the VMI experience with the mission of developing leaders of character to serve as citizen-soldiers. No one is sure how it works or exactly why it works, but the Institute certainly has produced more than its share of successful leaders.

On the flip side, VMI cadets are not perfect and like their civilian counterparts at other colleges and universities, they can swear and drink with the best of them. While drugs are often a part of the culture at other colleges and universities, they are not a part of the VMI culture; however, drug use does occasionally occur. One First Class cadet, who was in my class, was dismissed from VMI in his last semester. He, along with others found guilty of drug offenses, can agree to attend counseling while on suspension, and then return to VMI.

Leaders of Character

While VMI attempts to provide a loose framework of experiences for developing leaders of character, the model is being continually tested by cadets. For the VMI experience to be relevant, it must constantly adapt to a changing environment. For example, 9/11 was a wakeup call for VMI. At the time, the Institute was commissioning approximately one third of each year's senior class. This reality did not match the college's mission statement of producing citizen-soldiers who are trained to serve their nation in times of need. In 2008, the Institute commissioned 52 percent of its graduates and many cadets are currently serving tours with their Guard and reserve units in the Global War on Terrorism. When these young men and women return from deployment tours to complete their college education, they want and need to understand why VMI's rules and regulations are relevant to the world they experienced. Unlike their returning counterparts from World War II and the Korean War, today's veterans are not permitted to live off Post or to be married. Many of VMI's regulations and policies may seem "Mickey Mouse" to them, because not all the rules and regulations are relevant to "the real military." Still most return and provide mature leadership to the Corps.

These cadets and students at other colleges are the new generation. If VMI and other military colleges and service academies can adapt the learning environment in and out of the classroom to ensure relevance, these cadets will likely become the next "great generation." In many cases, these veterans are already leaders in the Corps and the military. They are individuals we can respect and admire, whether we are young or old. VMI and other military colleges have a responsibility to take these returning cadets to the next level and to ensure that they are not only warriors but leaders of character. For example, if a cadet is a Marine sergeant in the reserves and a combat veteran, why should he or she want to return as a private in the Corps of Cadets unless he/she is given a leadership role? The Institute, however, has determined that veterans will earn their cadet rank the same as every cadet. Some vets grumble that this isn't fair because they do not graduate with their Rat class, and it is therefore likely that they did not have the same opportunities to earn rank as their peers. On the other hand, many vets accept the Institute policy and move on to leadership positions such as a class officer or other key leadership roles. A veteran's positive example of selfless service goes a long way in the barracks and says a lot about his or her character and leadership. These young men and women are needed to provide a moral leadership example for their fellow cadets and troops.

Charles Cash (2010) was elected president of his class. Cash enlisted with the marines and served in Afghanistan, Iraq and Guantanamo, Cuba before enrolling at VMI. Cash said, "I tell everyone the Rat Line was very frustrating for me. . . It is hard to take being yelled at by a 19-year old kid after you have served your country in two combat deployments, but it was my decision to come here. So you have to just deal with it and look at the bigger picture, set an example for your Brother Rats and be a leader to them when the opportunity presents itself."[14]

Another example of the today's "great generation" of leaders is a 2007 VMI graduate. In April of 2007, I went to the new Gray-Minor baseball stadium. There, sitting on a front row seat, was a young man who looked older than the average senior. He was really enjoying himself and was busy heckling the umpire. His demeanor and bearing seemed to attract other cadets like a magnet. Cadets flocked to him as they came into the stadium. I noticed he had gone from sitting with the Corps Sergeant Major to being the center of a group of approximately 25 cadets or more. I later learned this young man was 25-year-old Sal Sferrazza from Long Island, New York. He was also the Corps Regimental Commander or First Captain (top dog) of the Corps of Cadets. Sferrazza told his fellow cadets that he was not in charge as First Captain because he had 1,300 bosses.

When I finally had the opportunity to meet Sferrazza, I clearly understood why the cadets were attracted to him that afternoon. He was charismatic, very positive and an extremis leader! Extremis leaders are those who are trained to provide "purpose, motivation, and direction to people when there is imminent danger, and where followers believe the leader's behavior will influence their physical well-being or survival."[15] Unlike crisis leaders, extremis leaders are self-selected.

Sferrazza joined the 106th New York Air National Guard rescue after graduating from high school in 2000. He went to SUNY Farmington to major in photography for one semester and was then sent to Lackland AFB for his basic training. While in the Air Force, Sferrazza made staff sergeant in record time and was the top graduate of the Airman Leadership School. Upon completion of Air Force training, Sferrazza returned to college in the fall of 2001. A few weeks later the Twin Towers of New York City were hit by a terrorist attack. Sferrazza claims the attack on 9/11 had a "dramatic effect on my life." He was immediately called to active duty and began 12 hour shifts, seven nights a week. He attempted to remain in college through the first semester of 2001, but was told not to register for the second semester as the unit would be deploying. The unit was first sent to Shaw AFB in Charleston, South Carolina for two months to help shore up an understaffed unit. In May of 2002, the unit deployed to provide security for a multinational, special operations base in Oman. During this time, Sferrazza began thinking about what he would be doing upon returning to New York

His mentor suggested he might consider attending a military college, and thus he applied to VMI while still in Oman. After completing his 110 day tour in Oman, Sergeant Sferrazza returned to New York and started visiting colleges. Sferrazza says he knew he wanted to attend VMI the second he passed through the entrance gates. While attending VMI, he attended every other New York Air Guard drill and made up his missed

Cadet First Captain Sal Sferrazza (2007)
a Air Force Combat Rescue Officer (CRO) leads by example.
Source: VMI Communications & Marketing

drills by working during holidays and furloughs. Sferrazza was also elected to serve VMI as the Honor Court Vice President.

In May of 2007, he graduated from VMI with a degree in International Studies and Arabic and no demerits. Upon graduation, he was commissioned a second lieutenant in the Air Force and selected for an elite 13-month training program, which prepares combat rescue officers (CRO) who serve in special operations units. To earn the spot, he competed in a grueling (physical, mental and psychological) one-week elimination against 18 Air Force officers, enlisted members, Air Force ROTC cadets, U.S. Army Rangers and other special operations members at Fairchild AFB in Washington. Sferrazza was the first VMI man while still a cadet to be selected for the CRO program. Five other VMI graduates have also cross-trained and completed the program. The mission of a CRO is to rescue downed pilots and crews of all branches. In the para rescue section (INDOC) of the course, 91 students started in Sferrazza's CRO class and 23 graduated including Gabe Hensley (2002).

On March 13, 2009 Sferrazza completed his nine year dream of becoming a Combat Rescue Officer. In addition, he graduated at the top of his class. Lieutenant Sferrazza said in an email, "Putting on my Beret and blousing my boots for the first time, I couldn't help but think about what it took to get to this point. The truth of the matter is; I didn't earn that Beret on my own. Standing on that stage I thought about every person that had taken the time to teach me. I thought about the people who stood by me while I chased this crazy dream."[16]

Another leader associated with VMI is Thomas "Stonewall" Jackson, a faculty member at VMI until the Civil War broke out. Jackson was a man considered by many as having impeccable character. Jackson was noted for his personal courage (Mexican and Civil War hero), Puritan habits, and honor. Once he even brought charges of immorality against a superior officer. A very religious man, he refused to use tobacco, alcohol, or profanity. He also gave up gambling and dancing when he joined a church in Lexington. His devotion extended to obeying the Sabbath. He even refused to write a letter if there was a chance it would be in transit on Sunday. He would discuss no secular business or read the newspaper. As a member of Lexington Presbyterian Church, he attended church every Sunday (although he often slept through services), organized a Sunday school class for young black slave boys and freedmen, and adopted the practice of tithing 10 percent of his income to the church.[17]

General George C. Marshall is another name associated with impeccable character. Marshall, VMI's most famous graduate, served as army chief of staff during World War II and later became secretary of state from 1947-1949 and secretary of defense during the Korean War (1951-1952). Brigadier General "Casey" Brower (ret.), former West Point department head and VMI dean, stated in an academic paper, "Marshall the soldier and his military career serve as a comforting reference point for thoughtful officers to guide upon when they feel they are in danger of losing their ethical and professional bearings… Statesman, as well as soldier, his character and accomplishments are so exceptional that he is regularly placed in the company of George Washington."[18] Upon Marshall's death on October 16, 1959, former President Harry Truman said, "He was the greatest general since Robert E. Lee. He was the greatest administrator since Thomas Jefferson. He was a man of honor, the man of truth, the man of the greatest ability. He was the greatest of our time."[19]

Hereclitus, an early Greek philosopher, once said that "A man's character is his fate." One of my favorite facts about Marshall drives home the importance of good character, because Marshall's character ultimately did define his fate. Marshall was known for being a straight shooter, someone who always spoke the truth and was not afraid of the consequences of being honest with his opinions. In World War I, he challenged General Pershing when he lashed out at Major Marshall's division commander, General William Sibert, for not preparing his troops for a trench warfare exercise. Marshall told Pershing that he was wrong and the problems rested at the division level. While Marshall's colleagues thought Marshall had ended his career, his actions had indeed sealed his fate. General Pershing liked his honesty and later made

Marshall his aide-de-camp upon returning to the states after the war.[20] This incident was a defining moment in Marshall's career.

Congress was used to Marshall testifying for hours on end without notes on the needs of our troops and how the war was going. President Franklin Roosevelt also became reliant on Marshall's assessment of the war and on his efforts to plan a war strategy devoid of political considerations and self-importance. Thus, when Roosevelt had to decide on who was to command D-Day forces, he selected Marshall to remain as chief of staff. Roosevelt felt no one was more knowledgeable of both the European and Pacific Theaters. He also had total trust in Marshall's judgment. Marshall never lobbied or pulled weight for the D-Day position he wanted above all others. He was an example of the quiet leader, one who gets the job done and does not blow his horn. Presidents, heads of state, and members of Congress knew he would not undercut them and make any secret deals. Leadership is built on trust. It is the foundation of our individual character and the greatest source of power that we draw from our followers we must faithfully serve.[21] George Marshall is a great example for all leaders and aspiring leaders.

In 2002 at West Point's annual National Conference on Ethics, I had the opportunity to sit next to General Fred M. Franks Jr., VII Corps commander during Operation Desert Storm. Franks co-authored *Into the Storm* with Tom Clancy. I was immediately impressed by General Franks. Trying to strike up a conversation, I asked him a question about whether he had plans to do news commentary leading up to Operation Iraq as many generals were doing. General Franks' response was direct and to the point. With a stern look he said to me, "I would never do anything that would be harmful to our troops." His words implied that armchair generals were aiding the enemy and putting their pocketbooks before their country. General Franks never appeared on TV during the war. His silence was a statement of his character.

In an interview with author Len Marrella, Franks had this to say about the importance of character:

Yes, character does count. Actually it is the foundation for everything you do in an organization especially a military organization wherein you may be asking your troops to put their lives on the line. Trust is the vital link and basic bond between a leader and his troops. Trust is earned continually by word and action. In the final analysis the ultimate alignment of the leader and the led is based on trust. When it comes to developing bonds of mutual trust and that being the basis for motivating people toward extraordinary performance, character in the leader is vital.[22]

Leading a Moral Life

Leading a moral life has long been an objective of the VMI education. With today's emphasis on political correctness, secularism, and separation of church and state, it may be difficult for some to identify with VMI's historical Christian roots. General Francis H. Smith, first superintendent of VMI (1838-1886), was considered by some to be a fanatical Christian. While Smith may or may not have been fanatical about his Christian beliefs, he did strive to establish a Christian environment, where cadets were groomed to be Southern gentlemen.[23] Smith even went as far as to conduct *Bible* classes in his office on Sunday afternoons and started the informal practice of presenting each senior a *Bible* at graduation. Today cadets are given a choice of books, but most select the traditional *Bible*.

To assist the commandant in maintaining the morale of cadets, the Institute today employs a full time chaplain, Colonel James Parks, and appoints a cadet chaplain to assist in campus services and to advise the chaplain on matters of morale and spirituality. Cadet religious organizations, such as the Fellowship of Christian Athletes (FCA) and the Officers Christian Fellowship, exist on Post so cadets can share their faith whether they are Christians or members of other religions such as the Jewish or Muslim faiths. These organizations are often organized by "Chap"

Parks to conduct "mission" trips. During spring break, groups often work in cities and regions where assistance is needed. One of my outstanding academic cadets (star man), Cadet Captain Eric Hunter, worked with a Habitat for Humanity crew in 2007 in New Orleans, and in 2008 he joined a group, which worked soup kitchens and fed the homeless of New York City. Eric is all about helping others achieve their full potential. Interestingly, Eric's role model as a Rat was First Captain Sal Sferrazza, who was profiled earlier.

While non-denominational services are held at the Jackson Memorial Hall (JMH), I have been impressed by the large number of cadets who attend church services in Lexington. One snowy Sunday morning, I found 30-40 VMI cadets and students from Washington & Lee at Lexington Presbyterian church. I found this rather astounding as Sunday mornings are the only time cadets can sleep in past normal reveille hours.

In his memoriam order of 1890, General E. W. Nichols, third VMI superintendent, spoke of Smith's influence on VMI, "General Smith aimed, as far as possible, to give to that government [Virginia], through military, a parental aspect. Hence he strove to make the control thorough, absolute, Christian. Such was the foundation upon which the founder of the school was builded (sic)."[24] This foundation was still strong through the 1900s and not until 2002 did the practice of saying a blessing before the evening meal end. A couple of cadets challenged the practice, and the case went all the way to the Supreme Court where the Institute lost.

VMI has another tradition tied to its Christian roots. After every athletic event the Corps and alumni sing the VMI Doxology (written by Cadet Croxton Gordon, 1904), sung to the tune of the Doxology sung in most Christian churches on Sundays.

The VMI Doxology

Red White and Yellow floats on high
The Institute shall never die
So now Keydets with one voice cry;
God bless our team and VMI

Especially in moments of battle, religion appears as a focus as it often is when one is faced with death. During my tour in Vietnam, on more than one occasion in the middle of the night, I found myself in a bunker praying along with other troops that we would not receive a direct mortar or rocket hit. It was I am sure for our chaplain, who was beside us in his flip flops and boxers, a defining moment for true believers.

Cadets also took their religious beliefs onto the battlefield. The four cannons displayed in front of the barracks at VMI were used by a Civil War battery led by Captain Pendleton of the Rockbridge Artillery. They are named Mathew, Mark, Luke and John in honor of the *Gospels of the New Testament*.

To honor VMI's brave cadets, Senator Henry A. DuPont sponsored reconstruction legislation to reimburse VMI for damage done by General Hunter. DuPont commanded Battery B, 5th U.S. Artillery, which bombarded the cadets at New Market. The cavalry escort commander was the father of Marshall's best friend, Ed Husted. The $100,000 in reparation funds was used to build the beautiful Jackson Memorial Hall Chapel.

The family of Lieutenant General George S. Patton Jr. has a long and storied history at VMI. George's grandfather commanded the 22nd Virginia Infantry at the Battle of New Market. George Patton Jr., who attended VMI his Rat year was a third generation cadet (father and grandfather graduated from VMI) and one of six Pattons to attend the Institute.

Lieutenant General George S. Patton Jr. said in a letter to his commanders in Europe on November 15, 1942,

"It is my firm conviction that the great success attending the hazardous operations carried out on

99

sea and on land by the Western Task Force could only have been possible through the intervention of Divine Providence manifested in many ways. Therefore, I should be pleased if, in so far as circumstances and conditions permit, our grateful thanks be expressed today in appropriate religious services." [25]

After World War II, the practice of marching Christian cadets to local churches for services was begun (attendance had been compulsory since Smith's superintendency with cadet companies rotating among churches) and remained mandatory until 1952. This practice remained voluntary until a 1972 Supreme Court decision. As a young child, my home was located on Letcher Avenue on the W&L campus. Cadets used the Letcher Avenue route to march into town for church services. On more than one occasion, cadets were pelted with beer cans by W&L fraternity members who were lodged in strategic locations at their frat houses along Jefferson Street and adjoining avenues. On at least one occasion, the cadets had had enough. They broke ranks and rushed the fraternity members. All and all, it was a good fight and is still a topic of war stories and campus lore.

The VMI Honor Code

In the early 2000s, I served several times as one of 16 facilitators for the West Point National Conference on Ethics. Approximately 150 college students from the service academies and other colleges and universities attended from across the U.S. Our opening session in 2000 was led by Dr. Donald McCabe, professor and founder of the Center for Academic Integrity at Rutgers University. He began with a report on the importance of honor and how cheating is a serious problem at American secondary schools and institutions of higher learning. He shared that studies show that academic cheating among high school and college students has risen dramatically over the last 50 years. McCabe says, "Students feel justified in what they are doing. They are cheating because they see others cheating and they think they are being unfairly disadvantaged. The only way many of them feel they can stay in the game, to get into the right school, is to cheat as well."[26]

Fortunately, VMI has an honor code, for there is no evidence to support that VMI cadets are any more honorable than other students when entering college. Assuming this is true, VMI can expect that less than 20 percent of its cadets have never cheated in high school. Bottom line math: VMI and all colleges have their work cut out for them as more and more students believe they are "entitled" to a degree versus fairly earning an education.

As a basis for discussion on the importance of developing honorable leaders, we need to examine the evidence stacked up against today's college students. One example and the most widely quoted is a survey done of A and B students nominated for Who's Who in American High Schools Students. It revealed the following astounding results.[27]

- 80% cheated
- More than 50% believe cheating is no big deal
- 95% say cheaters are not caught
- 40% cheated on a test or quiz
- 67% copied some else's homework

These results are carried over to college, graduate school, and later into careers. An interesting fact related to cheating is who is cheating. Today's cheater is likely to be the AP or honors student, who feels pressure from parents or others to get into the best college or graduate school or even imposes a self goal of graduating with honors.

The culture of cheating is also evident in the military school system. In late 2007, it was reported that the Army was investigating charges that enlisted soldiers were cheating on web-based tests required to earn points for promotions. Answers and tests were being shared on a soldier created web site called Shameschoo. com. General William Wallace, commander of TRADOC, says "Cheating violates our core Army values...Each and every one of the [soldiers] must live by Army values and leaders of character.

The institution depends on them."[28] Future NCOs like officers must be trusted by their followers.

In 2007, nine Duke University students who were caught cheating faced expulsion from Duke's Fuqua School of Business and 25 others faced lesser punishment for their roles in an exam cheating scandal. At the Indiana University Dental School, nine students were dismissed and 37 others given lesser punishments for cheating on an exam. What were they thinking? Can you fake doing a crown correctly or worse yet oral surgery? Maybe they'd just bill my insurance company for a more expensive procedure? [29]

The list goes on and on; however, it is probably safe to say that authorities did not directly learn of cheating violations from students. Why? Because today's student culture is opposed to "ratting out" fellow students. That includes not informing authorities that your friend has a fake ID, that an athlete has a no-show job, or that a drug dealer is operating on your dorm floor. The problem of not tolerating improper acts may improve on campuses as students start paying more attention to who is at the door and in the hall. Hopefully they will also notice who is looking at their paper. If nothing else, students should see the seriousness of not informing on other students. Validation of my point: the massacre of 31 students at Virginia Tech or at high schools when classmates knew something was in the works, but failed to report unusual behavior. Students need to be taught that it's okay to inform on other students, if for no other reason than it might save their lives. Unfortunately, reporting cheating or being a whistle blower is not always as easy as it may appear. According to the Josephson Institute of Ethics, less than two percent of all academic cheaters are ever caught.[30]

In the mid-1990s when VMI faced the possibility that women would be admitted to the Institute, a committee of administrators and faculty was appointed to outline the essential qualities of VMI. The number one item at the top of the list of qualities was the cadet-run honor system. The VMI Honor Code states that "A cadet will neither lie, cheat, steal nor tolerate those who do." This honor code is not that unusual. It is also found at several colleges including Virginia Tech, West Point, and the Air

Force Academy. It should be noted that studies indicate that schools and colleges with honor codes report significantly less cheating.[31]During my four years at Virginia Tech, I can honestly say, I never saw a student (cadet or civilian) cheat. Given that our tests and exams were never monitored, I must state the honor system was a very positive aspect of my Tech education. I graduated understanding the value of honor. Acting honorably becomes a habit or an attitude, not just a reaction to the fear of being caught. The same is true at VMI. I never monitored a test or exam, and students were allowed to move freely around the classroom buildings without asking permission. Why? At VMI, the faculty trust cadets to be honorable. At VMI character is as important as achievement.

So what makes the VMI honor system distinctive? The answer is twofold. One, for the most part, cadets believe the system works, and they believe in the ideal of honor as a treasured value. Second is the punishment given to those found guilty. VMI is the only "single-sanction" college for honor violations remaining in America. If a cadet is found guilty, he or she is immediately dismissed from the Institute. Cadets are not allowed to return to their rooms or to say goodbye to their friends and roommates or to collect their belongings. These items are gathered for them and presented to them as they are escorted from the Post. A "drum out" is later conducted after taps. The Corps is awakened and assembled by a drum roll. The Honor Court marches in a formation from Jackson Hall to the middle of the Old Barracks Courtyard. The president of the Honor Court then announces the name of the guilty party and states the charges before the Corps. A statement is then read, "He/She has placed personal gain above personal honor and has left the Institute in shame. His/Her name will never be mentioned in the four walls of Barracks again." Cadets then turn and return to their rooms.

As harsh as this ceremony may sound, in the "Old Corps" days at VMI and Virginia Tech, the cadet being drummed out of the Corps had to be present at the ceremony. As a Tech cadet I only witnessed one such "drumming out." These ceremonies were finally banned because they were deemed too humiliating to the cadet. The fact that civilian students at Virginia Tech were not drummed out may have also led to an end to the practice. One thing for sure, it certainly left a lasting impression on me. My honor was too important to squander for a few points on a quiz or test.

According to unofficial statements, VMI has approximately three to 12 honor violations a year that go to court. In recent years, most individuals who go to the court are found guilty. After a process of investigation and review by administrators and honor court members, cadet prosecutors today will first take a case to "pre-trial" to formally file charges. This process of investigation and review eliminates weak cases. Most cases are related to cheating; however, due to the high number of "official statements" that cadets are required to sign for PTs completed, etc., some cadets are accused of submitting false official statements. There are very few violations of stealing or toleration.

In my seven years at VMI, I was directly involved in only one honor violation. A Second Class cadet plagiarized eight pages of a sample marketing plan directly from the text and pasted it into a paper. Because of appeals and delays requested by the defendant, it took almost a full year before the case was heard. By then the cadet was a senior and probably hoping that his First Class status might affect the jury's decision. Upon learning he was being investigated, the cadet came to my office and admitted that he "might have gotten a little lazy." I think it would have been easier for all involved if he had admitted guilt to the Institute early in the investigation and resigned. Certainly his family would have saved a lot of time and money.

As a faculty member the case made me rather anxious. It was my responsibility to testify against the cadet. I had never attended a real court trial, and thus the seriousness of this situation made me uncomfortable. Knowing that my cadet would be represented by a real attorney, I had visions of *Perry Mason* and *Boston Legal*. If cadets cannot or choose not to be represented by an attorney, they may select a faculty member, classmate or whomever to assist them with their trial. Once the trial was over, I left with a feeling that the cadet's rights were protected and that he certainly got a fair trial. After all, he was judged by a court of his peers (non-court members). Unfortunately, all cadets do not share the feeling that the Honor System is administered fairly, but a large majority does believe in the system.

The court is presided over by a senior president and composed of 12 members, all seconds (six juniors) and firsts (six seniors) elected by their peers. The First Class members will serve as president of the honor court, prosecutors, investigators, and defense and education members. The six Second Classmen serve as assistant prosecutors. It is a high honor to be selected to serve the court and each cadet is interviewed before being accepted. The interview determines if cadets have the time to commit to this extra duty. Being on the court is very demanding and a huge responsibility. The court meets six days a week at 2300 after taps. Duties could keep them up way into the night.

Election to the court is not a popularity contest. Those elected are individuals who have demonstrated they can be trusted to protect the rights of the cadets and to uphold the integrity of the honor system. Once the trial is completed, the three faculty honor court representatives meet with the superintendent, who always asks if proper procedures were followed. In addition, this group of faculty provides guidance to the court, education to the faculty and staff, and act as the liaison between the superintendent and the Board of Visitors.

For a comparison to VMI, one can examine recent cases at the service academies. All service academies have experienced one or more cheating scandals over the years. In the "Old Corps" days, any cadet would have been kicked out of an academy if found guilty of an honor violation. Since the 1990's the superintendents of the academies have played a greater role in the review of recommended punishments and are thus now playing the role of an appeals court. Cadets found guilty can be allowed to stay on honor's probation or can be suspended and allowed to return at a later time.

In a 2007 cheating scandal at the Air Force Academy, the Cadet Honor Court investigated 40 fourth-class cadets (freshman) accused of cheating on a weekly knowledge test. The test compromise was reported by other cadets, who did not want to violate the school's no tolerance clause in the honor code. "Of the 40 cases, 29 cadets admitted cheating, five denied cheating, and six cases concluded the cadets did not cheat. The five who denied cheating met Wing Honor Boards in which two were found guilty of violating the Honor Code. A total of 31 cadets were found in violation of the Cadet Honor Code. One cadet who admitted cheating resigned."[31]

The case was reviewed by the commandant of cadets who determined that 12 of the violators would be placed on honor probation for six months and the remaining 18 should be dismissed. Two cadets recommended for disenrollment resigned. Nine of the remaining 16 appealed their punishment to the academy superintendent, who upheld eight of the nine recommendations.

Moral and Ethical Development

A focus of this chapter is to understand how VMI develops leaders from a moral and ethical perspective. Thus far, we have examined the value of character and the importance of honesty and truth as foundations for leadership. Now we must examine how individuals develop morally and to examine the role VMI plays in this development. According to social psychologist and author of *The Psychology of Moral Development*, Lawrence Kohlberg, there are three levels of moral development.[32] VMI has no specific step-by-step guide for moving from one stage to the next. Cadets will arrive at the next level when they see the difference that moral decision making is accomplishing in their personal lives. The Institute does attempt through a series of classes on honors education to teach cadets that there is more to decision making than just obeying rules and pursuing their personal interests. Cadets are encouraged to discuss these issues and engage in debate the merits of moral leadership.

In the first levels, the cadet is aware of cultural prescriptions for right and wrong. His or her response when faced with a moral dilemma will be based on two stages: 1) How can I avoid punishment? 2) Will I be helped with a higher grade, avoid punishment, or will things actually be more pleasant? At this stage, physical consequences often determine the behavior. For example, it may be a lot easier for a Rat to lie about having a cell phone, than to admit it and thus walk hours of penalty tours.

A second level is based on the approval of others (stage three) and compliance with authority (stage four). As a stage three "Rat," my fellow rats and I at Virginia Tech did some really stupid things to earn the approval of upper-class cadets. As a result we were put on social probation for six weeks. Social probation amounted to room confinement and studying with the door open and in a "braced" (at attention) sitting position (front two inches of the chair). It certainly would have been smarter of us to gain the approval of our cadre and the commandant's staff, but then we were Rats and we were supposed to be stupid. The same can be said for my VMI cadets. Rather than have me remind them every day or force me to "bone" (report) them, it would be much smarter if they tried to gain my approval by following the regulations such as not wearing their covers (hats) in the academic building or bringing coffee into the classroom.

At VMI, negative behavior for the approval of others has ranged from buying alcohol for minors to the occasional rebellion (burning mattresses and breaking furniture), food fights, or marching, without authorization, as a unit into town as a protest. Keep in mind you do some of these things whether you want to or not, because the class hierarchy has decided it. Not to go along could create serious social problems. While none of the above are considered honor violations at VMI, they are considered violations of the *Blue Book* or cadet administrative regulations. A few cadets love to beat the *Blue Book*. They like to discover how far they can go without being "called out" and nailed with demerits. Even George Marshall used to "run the block" (AWOL) after taps to court his first wife. Obviously, cadets who are intentionally disobeying Corps regulations have not achieved stage four where obedience to authority is an expected behavior. They have not learned to comply with orders, because it is wrong not to obey. On the positive side, most cadets have advanced to step four by the time they graduate.

Finally, at stage five of the third level, cadets begin to demonstrate social ethics. They participate in community discussions and are open to listening to other views of opinion. For example, a cadet might be opposed to the war in Iraq, but he or she would not disrupt Senator McCain when he made his strong speech for the war in April of 2007 at VMI. The cadet would uphold McCain's right to express his opinion. As another example, the cadet accepts that when there is a difference in individual views and group interests, the majority rules. Thus, when a roommate is found guilty of an honor violation and a cadet believes his BR innocent, the cadet must accept the decision of the Honor Court. It is no different than accepting that our favorite political candidate did not win an election or that the Supreme Court voted in opposition to our opinions.

In stage six of the third level, cadets can make decisions without external forces compelling them to act in a way that they consider morally wrong. Choice is made as a matter of individual conscience and personal responsibility. Each cadet accepts personal accountability for his or her actions. No effort is made to blame the system or others.

Evan Offstein, a former VMI assistant commandant, recommends in his book, *Stand Your Ground*, what he calls the "Three Rules of Thumb."[33] These rules are actually questions he learned as a cadet at West Point. Offstein says leaders should ask the following:

- Rule One: Does the action attempt to deceive anyone or allow anyone to be deceived?
- Rule Two: Does this action gain or allow the gain of privilege or advantage to which I or someone else would otherwise be entitled?
- Rule Three: Would I be satisfied by the outcome if I were on the receiving end of this action?

Offstein believes in "test driving" these rules. If any rule is broken then you know that in your moment of truth, you will face a moral dilemma meltdown. In his opinion, when in doubt, we should always take the high ground. I have always found this to be the best policy when dealing with gray area issues, and I highly recommend this stance to my cadets when discussing ethical issues.

Conclusion

The Character Education Partnership (CEP) promotes good character by creating forums on good character for discussion amongst educators and by recognizing national schools of character. CEP has identified 11 principles for effective character education, some of which do not apply to VMI (not K-12).[34] Of the eleven key principles identified, VMI actively promotes seven of these to help ensure that VMI graduates are leaders of good character:

1. Promotes core ethical values as the basis for good character.
2. Defines character comprehensively to include critical thinking, feeling, and behavior.
3. Uses a comprehensive, intentional, proactive, and effective approach.
4. Creates a caring school community of professors, administrators, staff and alumni.
5. Provides a meaningful and challenging curriculum that helps all students to succeed.
6. Fosters students' intrinsic motivation to learn and be good people.
7. Fosters moral leadership.

CHALLENGING THE MIND
INTELLECTUAL & PROFESSIONAL DEVELOPMENT

*"Education is what survives when what has
been learnt has been forgotten."*

B. F. Skinner, American psychologist

Importance of Intelligence in Leadership

The foundation of every college or university is its
academic programs. Colleges may have great athletic teams
for which they are best known, but the primary purpose of any
institute of higher learning is the intellectual and professional
development of its students. This is accomplished through a
variety of academic programs, which enroll students in majors,
minors, concentrations, or fields of study. As a college professor, it
is my responsibility to enhance students' "all-around effectiveness
in activities directed by thought."[1] VMI, while offering similar
academic programs as most colleges, goes a step beyond most
academic institutions by constantly exposing our students to
problem solving and decision making experiences that go much
further than the strict routine expected in an environment of
military regimentation and discipline. At VMI the classroom
environment is called the "academic ratline" – to many cadets it
is equivalent to the regimental Ratline.

The evidence of an academic program that values
the importance of intelligence and an all-around approach to
learning abounds at VMI. Dating back to 1921, when Samuel
Washington, Jr. was selected as VMI's first Rhodes winner,
the Institute has produced eleven Rhodes Scholars. The
Rhodes scholarship is the world's oldest and most prestigious
international study award. Rhodes Scholars are selected on the
basis of high academic achievement, integrity of character, spirit
of unselfishness, respect for others, potential for leadership, and

physical vigor – all characteristics in common with the qualities aspired to by the VMI citizen-soldier.

I had the pleasure of working with/for two of VMI's Rhodes Scholars, Brigadier General Lee D. Badgett (1961) and Lieutenant General Josiah Bunting III (1963). From their days as cadets, Badgett went on to have a distinguished career in the Air Force and academia (former academic dean at VMI and several other colleges), and Bunting rose to prominence as a successful author and three-time college president (including superintendent of VMI). One of the Institute's most recent Rhodes Scholar is Michael N. Lokale, class of 2003 and a citizen of Kenya. At VMI Lokale was a biology major (3.784 GPA), a Regimental staff officer, and a two time All-Southern Conference champion (in 400 and 800 meter races) as a member of both the VMI track and cross-country teams. Lokale attended the University of Oxford for graduate school and is currently attending the Virginia Osteopathic Medical College at Virginia Tech training to become a physician. Lieutenant General Bunting predicted Lokale's potential early in his cadetship when he said, "Michael is the outstanding VMI cadet of my time here. He has a first class mind and a first class temperament."[2]

In 2008, cadet Gregory Lippiatt (2009) was named VMI's 11th Rhodes scholar. Lippiatt, a history and English major, was one of 32 scholars (1,500 applied) selected for the honor. During his tenure at VMI, he was a regimental commander, member of the Rugby club, an actor in several VMI plays, and the editor of the *Sounding Brass* (Institute literary magazine). After completing his degree at Oxford and serving as an officer in the Army, Lippiatt, plans to complete a doctorate in medieval studies.

Another recent cadet, Matthew Sharpe, was a Rhodes and Marshall finalist in the 2004 competitions. Sharpe was a member of the Regimental staff (S-5), Institute Honors Program, and the football team (All-Big South first team player). I was fortunate enough to teach Matt in my Business Leadership course. I have been asked several times what it is like to have a student of that caliber in class. Perhaps, only a teacher can appreciate my response. "He was always prepared." It didn't matter if the football team had

VMI's newest Rhodes Scholar, Gregory Lippiatt (2009),
discusses his goals with Louis H. Blair,
the Mary Moody-Northen Chair in Arts and Sciences.
Source: VMI Communications & Marketing

been away on a trip, and he had been on a bus or plane just hours before class. Sharpe was also selected as National Scholar-Athlete of the Year by the National Association of Collegiate Directors of Athletics (NACDA). Matt, a computer science major with a 3.91 average, was well-rounded like most Rhodes finalists.

Matt wore his character on his sleeve and possessed an intellectual curiosity that few students possess. I specifically remember one discussion after class, when he reflected that a high school classmate and close friend told Matt that he had made a terrible decision to attend VMI. Sharpe told me that he never regretted his choice of colleges and that he had lost respect for his former classmate who couldn't see the value in selecting a challenging and rigorous college. Matt's hard work paid off for him. He won a full ride to Carnegie-Mellon, where he studied computer science as a graduate student.

Another student athlete with great intelligence was Kelly Sweppheniser (2006). Kelly, another student in the Economics and Business program, was a quiet leader on and off the field. Besides earning many academic honors, he was drafted by the Toronto Blue Jays and is currently playing minor league baseball. While at VMI, Kelly was First Team Big South (third-baseman) as both a junior and senior and was awarded the Big South's 2006 Scholar-Athlete of the Year Award. In 2003 he was named a Louisville Slugger Freshman All-American and a member of the 2006 *ESPN The Magazine* Academic All-District III team. "In his senior year, he led the Keydets with a 350 batting average, 72 hits, seven home runs, and 41 RBIs."[3]

Intelligence is a very general trait of mental capacity. Intelligence is a building block in a pyramid of factors such as personality traits and preferences, values, interests, motives, and goals. It involves the "ability to reason, plan, solve problems, think abstractly, comprehend complex ideas, learn quickly and learn from experience."[4] Research shows that it is difficult to change intelligence, but not impossible. Intelligence can be modified through education and experience. It also does not affect behavior equally across all situations. One's environment and genetic heritage play an important role in creating differences in intelligence. Unlike character, personality, and creativity, intelligence can be measured very accurately (both reliable and valid results).[5]

Robert Sternberg's work on intelligence is considered some of the most comprehensive and compelling of the past 20 years.[6] His triarchic theory of intelligence offers significant implications for leadership. According to his theory there are three basic forms of intelligence: analytic intelligence, practical intelligence, and creative intelligence. These three types of intelligence correlate well with the three types of leaders discussed in Chapter 3; however, it should be noted that while leaders may possess all three types (multidimensional) of intelligence, they may be stronger in one area than another. Furthermore, different types of situations require different types of intelligence. While most leaders are intelligent, they may be weaker in some types of intellectual skills than others.

Analytical intelligence allows leaders to be quick learners. It is closely associated with a hot button in education identified as critical thinking. Leaders possessing a good analytical intelligence have a foundation of knowledge to draw on and can develop the necessary connections needed to make deductions, assumptions, and inferences with unfamiliar information and situations. Cadets with good analytical skills often perform well in conducting field training exercises or leadership reaction courses. Analytical intelligence has been shown to be the best indicator of job performance.

Practical intelligence is better known as "street smarts." It involves "knowing how things get done and how to do them."[7] For example, cadets who have prior military experience have practical intelligence when it comes to military inspections, medical emergencies, and physical training. They can usually lead other cadets and are better prepared to handle underperforming Rats and other classmates in their company. In general, upper-class cadets also possess practical intelligence. Through their daily routine of classes, drills, parades, inspections, physical training, and extra-curricular activities, they learn through experience how and when they can take short cuts within the system. This in turn helps them optimize their time and make their lives a little simpler.

The final component of triarchic theory, creative intelligence, provides the skills to produce work that is both novel and useful. A rapidly changing world demands that 21st century college graduates possess a high degree of flexibility and adaptability. "These skills are often embodied in the ability to think both critically and creatively."[8] Creativity according to Lee J. Cronbach is the ability to make fresh observations and to see things in a different and appropriate light.[9] The triarchic theory also includes evaluation. For example, cadets in my marketing management course are sometimes asked to critique the marketing plan of a local firm and to make suggestions for how to improve the plan. Cadets in a creative writing course may be asked to evaluate a classmate's paper and to give suggestions for improvement. Or in an advanced engineering class, cadets are asked to develop a new product, apply for a patent, and then work with a business entrepreneurship class to develop an effective business plan.

Leaders and Stress

While intelligence plays an important role in problem solving and decision making in some situations, it can also be a disadvantage in others. Recent research suggests that the effectiveness of a leader will often depend on whether or not the leader is in a stressful situation. When placed in a stressful situation, some leaders simply "freak out" and lose control of their emotions. Once this occurs they are no longer effective leaders.

Cognitive resources theory examines the relationships between leader intelligence and experience levels and group performance in stressful versus non-stressful conditions.[10] VMI prides itself in providing an education under "adversarial" conditions. The idea is to give cadets stressful situations where they will often fail, when placed in conflicts. For example, a cadet may have to choose whether to spend time studying for a test or preparing his or her room for inspection. Knowing they have a no-win situation can cause tremendous stress regardless of which decision they make. The cadet wants a good grade, but doesn't want to spend his or her spare time walking penalty tours for failing room inspection. Because of the interpersonal academic stress, a cadet can become so focused on test performance that he or she may fail to perform at an all-around optimal level. I have seen a few cases at military colleges where some cadets actually choose room preparation or athletic training over academic preparation. Why? The likely answer is that military or athletic status within the corps is more important and personally rewarding than academic performance.

On the flip side, I've also witnessed many students become so obsessed with their grades that they fail to learn much of anything. Grades become the mantra versus obtaining a solid education. Good grades and true learning are earned through continuous preparation rather than overnight memorization of facts, which are quickly forgotten. As educators we can increase student stress by providing fewer but higher value tests with more questions, or we can lessen the stress by increasing the number of lower value tests and giving fewer questions at a time. Students are less likely to become stressed and learn more, if we test them more frequently. If anything, it is imperative that students over learn knowledge and skills through a series of learning experiences. Even very bright students with high IQs usually perform at higher levels if they develop solid study skills coupled with learning experiences over an extended time.

Two stressful situations, which affect academics at VMI, are physical exhaustion and a lack of sleep. End of semester and mid-term exams are also stressful times. Because some cadets attempt to "be all, do all," they may fall into a trap of staying up half or at times all night to accomplish their many academic, military and extracurricular tasks. The best students are masters of time management and never seem to be stressed by the day's activities. Others, having failed to keep up and effectively manage their time, will show up for class and immediately go into a sleep mode. The stress of trying to do too much or ineffectively managing their time eventually catches up with them.

Initially, I was offended by cadets sleeping in class; then I came to the realization that at a civilian college, these students simply wouldn't attend class. At VMI, only those on the Dean's List (3.0 or higher), have a documented illness, assigned honor court ("Institute business") or guard duty, or are athletes on game days can miss class if excused. Any cadet, for any reason accruing nine or more absences is automatically dropped from class. Finally, unlike so many colleges today, VMI requires that all professors give a final, comprehensive exam. The exam should be weighted no less than 30 and no more 50 percent of the final course grade. A poor performance can easily bring a grade down by two letters.

Generally, I tell cadets that unless you have a high F, don't expect your grade to improve. Cadets (or any students) seldom improve their letter grades as a result of a final exam; especially at VMI where we do not use a grading scale of pluses and minuses as so many colleges do. For example, a C is worth 2.0 points, while at schools on a plus/minus system, the grade could range from a 1.75 to a 2.5. While there are exceptions, exams are more about grade maintenance than learning a semester's worth of work in three exam hours. Regardless, exam time is always a stressful time for cadets, who must manage an exam schedule that may include as many as two exams in one day. These situations test their intelligence and provide stress under fire ... all experiences needed to be successful leaders.

Academic Programs

"The heart of the Institute's academic program - indeed, infusing the VMI education in its broadest sense – is leadership development."[11] It is important as an institute of higher learning that we provide cadets every opportunity to develop each type of intelligence needed by tomorrow's leaders. In 2001, Brigadier General (ret.) "Casey" Brower became academic dean at VMI. Brower came from West Point, where he retired as department head for Behavioral Sciences and Leadership. With the superintendent's and Board of Visitors' approval, he followed the new Institute master plan (Vision 2039) with its focus on leadership. In addition, Brower strongly encouraged the development and creation of new academic programs and standards to improve academic excellence at VMI and to ensure that the Institute became not just a good college but rather one of America's "premier" undergraduate academic colleges. By the spring of 2002, VMI was named the number one public liberal arts college in America by *U.S. News & World Report.*

"Vision 2039 helps ensure that VMI graduates will be broadly educated, adaptive leaders of character who contribute to their communities, their nation, and the world. A major academic goal is to ensure that every cadet, regardless of academic major, is properly prepared for leadership and citizenship in the 21st century."[12]

A key weapon in the dean's academic arsenal is the Jackson-Hope Fund. The endowment was created in 2001 as a legacy of the Jackson-Hope Medals given each year as VMI's highest academic awards. The medals' program was started in 1876 by a group of Englishmen who admired Thomas "Stonewall" Jackson. The Honorable A. J. B. Beresford Hope, a member of the British Parliament, authorized the establishment of the fund as a memorial to General Jackson. The endowment goal for the fund is $50 million; in the first six years, the fund's overseers dispersed $7 million in grants to support academic opportunities for faculty and cadets. From a faculty perspective, I can testify that the fund truly makes a difference in the academic excellence of the Institute.

Academic Instruction

VMI was created as an experiment in higher education. A VMI education was designed to provide practical courses in engineering, industrial chemistry and other technical courses. The curriculum was in sharp contrast to the more abstract and traditional college curriculums found at other schools. VMI was modeled after West Point and influenced by the French technical school L'Ecole Polytechnique, from which VMI's first Board of Visitors president, Colonel Claudius Crozet, graduated. The Institute was the first state supported military college in America. Those associated with VMI will proudly tell you that the Institute is different from West Point, where the mission is to prepare leaders of character to serve the nation as officers in the U.S. Army. VMI's mission is to develop citizen-soldiers who are leaders in all walks of life.

From its initial beginnings, VMI adopted West Point's Thayer method of instruction; however, over time, like the academy, the method has been modified and now is used on a limited basis. The Thayer method, named after West Point's first superintendent, Sylvanus Thayer, required small classes and daily recitation using the blackboard. Today VMI still has small classes (1:13 ratio of faculty to students). The largest class I've experienced at VMI is 24 and the smallest was four students. Normally class size ranges from 12 to 20. Small classes allow professors to quickly know their students. Interaction between the professor and cadet is common in and out of the classroom. It is very common for faculty to entertain cadets in their homes, know their parents, and dine with them on and off campus.

I taught Third Class cadets the principles of marketing course, which is mandatory for all students in the Economics and Business major. Every cadet in our major must take the course I taught; thus, I am exposed to every cadet before their Second Class (junior) year. When they enrolled in my advanced marketing and business leadership electives in their last two years, I already had a rapport with them and was aware of their abilities and interests. Prior to coming to VMI, I had rarely experienced this much direct exposure to students. With the exception of a few expensive private colleges, this benefit does not exist at most public colleges.

The initial curriculum at VMI was not very comprehensive. It was composed of math, mechanics, chemistry, engineering, tactics, French and German languages, and French, German, and English literature. Until 1848 cadets could graduate in three years, but after 1848 the curriculum required four years. The first four-year class began in 1849. In 2006-2007 the Institute introduced a new 47-49 hour core curriculum. The purpose of the Core Curriculum is "to develop foundational knowledge and skills that are essential to VMI academic and military mission."

One method of creating more in-depth exposure to students are the faculty & cadet lunches.
Pictured here left to right are: Col. Bob Moreschi, Cadet Trevor Stanco, & Col. Cliff West.

Third Class Cadets (sophomores) alert and ready for a business class.

The Core Curriculum includes 15 hours of Citizenship. According to the VMI Quality Enhancement Plan, "This component of the curriculum core provides classroom and practical experience with the rights and responsibilities of citizenship in a democratic society and prepares cadets for effective leadership and service to the nation, whether in the military or in another profession of their choice. It also equips them for personal success in the community of the Corps of Cadets and in their early life after graduation."[13] It should be noted that the Core Curriculum requires all cadets take 12 hours of ROTC (Army, Navy/Marine, or Air Force) throughout their entire four years at VMI. Furthermore, a new 3-hour interdisciplinary course in Leadership and two, 2-semester pass/fail seminars, Success at VMI (Fourth-Class year) and Success for Life (Third-and First-Class years) have been added to the curriculum.

It may surprise you that all cadets are also required to take boxing. You may be the smartest male or female in the barracks, but the boxing ring has challenged many a cadet over the years who hoped to achieve the coveted and rare 4.0 average for their four years at VMI. Everyone in the Corps of Cadets knows what course created a 3.96 average when Institute honors are announced at the end of the year. In addition, every cadet must pass a swimming course, which requires survival skills such as drown proofing.

The remaining components of the core curriculum consist of 12 hours of World History (6 hours) and Civilizations (6). The purpose of this component is to encourage "cadets to contemplate and think critically about major challenges, accomplishments, and errors of humanity on a global scale. The goal is to cultivate and energize cadets' capacity for self-examination through analysis of the products of civilization."[14] Furthermore, an integrative experience consisting of two Writing-Intensive courses (one must be taken in their major) and a capstone course must be included in the cadets' major.

Capstone courses must be culminating experiences. For example, cadets majoring in Economics and Business take Business Policy as their capstone course for the principles and concepts of business and economics. In this course cadets take the Educational Testing Service (ETS) national field test (over 400 colleges annually participate) on business content. Since first taking the exam in 1992, Economics and Business cadets have ranked in the top 13 percent in the nation. Since instituting a major curriculum revision in 2003 for the Economics and Business major, VMI has ranked in the top 5 percent overall each year and in the top one percentile in economics. This record is truly phenomenal, and few if any colleges can match VMI's 15-year record.

Colonel Stewart Husted, The John and Jane Roberts Chair
of Free Enterprise Business, presents Business Week/Irwin 2004 Student
Case Writing Competition Award to Cadet Kyle Clark.
Cadet Andy Karnes was also a co-winner of the competition.
The team finished in the top 9 entries
nationally and earned an Honorable Mention.

International Study

A number of cadets take advantage each year of the study
abroad program, either in conjunction with a summer session or
during a semester abroad. The Institute has programs established
with such universities as Cambridge (Britain), Oxford (Britain),
St. Andrews (Scotland), Freie University (Germany), University
of Waikato (New Zealand), and American University in Cairo
(Egypt). The VMI summer session includes study abroad trips to
Peru, Hungary, Japan, Morocco, and England. In addition, cadets
have the unique opportunity to be selected for an international
military academy exchange. These cadet exchanges include

The Lithuanian Military Academy, Saint Cyr (French Military Academy), The Bundeswehr University (Germany), L'Ecole Polytechnique (France), and The Hungarian Military Academy. Cadets participating in these opportunities must have a minimum 2.5 GPA.

Cadet Laird Bryant (2010) learned firsthand how French cadets train for military action during a two-week exercise in mountain warfare in the French Alps. Bryant, an exchange student for a semester at Saint-Cyr academy, received training in snowshoeing, rock and ice climbing, bouldering, and rappelling with packs and weapons. A frequent exercise drill included searching for possible avalanche victims through use of radio transmitters, which each cadet wore. The cadets also got to participate in a simulated attack on an old fort in an area of France sprinkled with 15 other forts ranging back four centuries.[15]

Internships

The internship program at VMI is administered by each individual department. From my experience, most of the internships completed by cadets are not the run-of-the-mill average internship completed at many colleges and universities. For example, the standard at many colleges is 48 hours of work equals one hour of academic credit. In the Economics and Business Department, cadets are required to work over 300 hours for a three-hour course, read a related book of their choice, complete a research paper, and present a summary of their research paper and internship to the faculty. Many of our cadets are offered jobs a year in advance of graduation by their sponsoring companies. Corporations such as Booze-Allen, Ascenture, Raytheon, and Long-Foster Real Estate Development, to name but a few, include positions around the globe. Cadet work varies from performing accounting audits, consulting, developing business plans, researching patents, lab work, and business proposals.

Cadet Joshua Gauthier, an intern at Raytheon in 2005, said "What I go through here at VMI has prepared me really well for life in the real world. Learning from both leadership as well as how to be a subordinate is extremely important. Knowing how to handle all different types of situations and people is instrumental in getting ahead, and I feel VMI has given me tools to do this."[16]

Faculty

VMI has a long history of esteemed and popular faculty. To put it simply, VMI is blessed with many outstanding professors. Student evaluations of faculty on average are significantly higher than most colleges. I attribute this fact to the complete dedication and commitment of so many faculty to our students. For example, many faculty keep evening office hours when cadets are most likely to have free time. One professor, Colonel Francis Bush, is legendary for his late-night hours and weekends he dedicates to students. Cadets in the class of 1998 recognized his commitment by making him an honorary class member. Another faculty member is also in a league of her own. During our Google Ad Campaign project for my marketing course, Major Elizabeth Baker kept office hours on Friday nights from 7 PM to 10 PM.

More than likely you will find the names of legendary faculty members on Post buildings. Post academic buildings are named for professors Carroll, Maury, Jackson, and Mallory. At other colleges, building names often reflect wealthy donors. At VMI they reflect the special place faculty have in the lives of cadets. At one nearby college, an academic building was named for a couple, who gave an estimated $2-3 million over a life time. At VMI, naming opportunities for donors have been limited to athletic facilities such as the recent construction of the Gray-Minor Stadium (baseball) and the P. Wesley Foster Stadium (football).

What you won't find is a list of the nicknames given to professors. One well-respected and tough professor, Robert P. "Doc" Carroll arrived in 1928 as an assistant professor of biology. "Doc" Carroll headed up the premed program and was well known by regional medical schools. If Carroll gave his approval, cadets knew they had a passport to admission. The department head of economics and business (1977- present), Floyd Duncan, earned the nickname "Flunkin" Duncan. Needless to say Duncan is well known for his high standards. Francis "Monk" Mallory was head of the physics department and is reported to have taught more cadets than any other man (1886-1940). That record may be surpassed soon by Colonel William Badgett (1953) in the English department who never left VMI after returning from earning his masters from Harvard University. In 2008, Colonel Badgett had served VMI 53 years teaching English and fine arts courses. As mentioned in Chapter 1, other well-known faculty include Commodore Matthew Maury, George S. Patton II (father of General George Patton), John M. Brooke, and General G. W. Custis Lee (eldest son of Robert L. Lee).

More recently, Dr. Mohamed Taifi, an instructor of Arabic and French, was honored in 2005 by the Moroccan newspaper *Imazighen* as the Moroccan Man of the Year for his work in education and as the permanent representative of Morocco to state assembles in Quebec, Canada, Paris, France and other international locations. Under Dr. Taifi's leadership, the Arabic program has grown since 2002 from two part-time instructors and one course to three instructors and nine courses in 2007.[17] The program has gained much attention and is a popular course with cadets interested in international studies, homeland security, and military careers.

In 2004 and 2005, two VMI engineering faculty were selected by the Virginia State Council for Higher Education (SCHEV) as the "Rising Star" faculty member for the state of Virginia. Only one junior faculty member is selected each year from a competition of nearly 60 public and private college/ university nominees. Lieutenant Colonel James C. Squire, an associate professor of electrical and computer engineering won the first award given in 2004. Major Matthew R. Hyre, an assistant professor of mechanical engineering, won in 2005. Hyre at the time of the award had worked with more than 25 cadets in his research. Eight cadets co-authored with Hyre and one shares a patent with him. Brigadier General Brower, the academic dean said of Hyre, "Matt's love of teaching and commitment to his students have captured attention across our campus. His classroom models organization, imagination, and effectiveness, and students flock to him."[18] Both Squire and Hyre are graduates of the United States Military Academy and have Ph.D.s from MIT.

In 2008, Dr. Duncan Richter received SHEV's Outstanding Faculty Award, the highest award given by the Commonwealth to honor teaching, research, and public service at Virginia's public and private universities. Richter, a professor of philosophy and a British graduate of Oxford and the University of Virginia, taught at VMI for 12 years prior to winning the award. During his tenure he also won VMI's Thomas Jefferson Teaching Award in 1998 and the Matthew F. Maury Award for Faculty Research. Dr. Richter has a reputation for his profound and objective thinking and for stretching cadets' minds with rigorous courses. Cadets say he is exceptionally engaging and that he is one of those teachers who you rarely want to disappoint.

During my first year as the Wachtmeister Visiting Chair in Science and Technology, I had the opportunity to observe the faculty as a non-tenured member. After my bonding experience at Eagle's Landing, I began my "Rat" faculty year. Everyone in my department was extremely helpful in assisting me make the necessary adjustments, which just so happened included being updated on military customs and uniform and insignia wear. Unless you are an adjunct/visiting professor or a non-U.S. citizen, you are expected to wear a military uniform. Applicants for U.S. citizenship are also allowed to wear the uniform. Wearing the faculty uniform correctly is very important at VMI. Faculty set the example for cadets, who at times can be lax on appearance standards. It is much easier to wear dirty white ducks to class, if a faculty member habitually wears un-pressed shirts, pants, and un-shined or unauthorized shoes. Faculty are the academic leaders on Post and must set the standard for cadets in the classroom.

Military customs may not seem like a big deal, but to an outsider or cadet they can be intimidating. For example, because we all wear a military officer's uniform (usually Army unless the faculty member has served in another branch), cadets are required to salute us. One new faculty member on his first day got a little carried away with the saluting requirement and was stopping every cadet in sight who did not salute him. No one told him that the custom is to salute when someone is approximately six to 30 paces from you, not 50 yards away. Faculty are also required to salute other faculty and staff who outrank them. For someone unfamiliar with military rank and poor eye sight, this can be an anxiety moment. I hate to admit this, but after a 28-year Army Reserve career, I still have a hard time distinguishing Navy rank. At my age and rank, I just figure that they'll probably have to salute me. Presently there are no admirals on post; however, in 2007-2008 there were five generals, one of whom is a retired four-star.

Academic Advising

Unlike faculty members at other colleges, VMI faculty are evaluated annually on four (versus the normal three) key components: teaching, scholarship, service, and cadet development. Excellent teaching is by far the most important component and is weighted heavily for retention, promotion, tenure and merit pay considerations. Cadet development includes all aspects of contact with cadets that enhance cadet growth outside the academic classroom. A major element of cadet development is academic advising. With the exception of the service academies, I have rarely witnessed a more comprehensive approach to academic advising as I have at VMI. To offer a comparison of VMI to other colleges and universities, an examination of the 2007 National Survey of Student Engagement, provides evidence of VMI's advising excellence. The VMI Fourth Class cadets rated their advising experience a 3.45 (4 is excellent and the highest score possible). Freshman at other colleges ranked their advising experience a 2.98. As students and cadets advance through college, they generally need less advising, and consequently the ratings tend to drop. The average for advising by VMI First Class cadets was a 3.14, compared to the national average of 2.84 by seniors.

Major James Bang advises a student about his class schedule.

In many VMI departments, new faculty members are usually not assigned advisees in the first year; however, the faculty member will normally observe and assist his or her mentor. In my second year, I was assigned six Rats and an assorted collection of upperclassmen. The advisor's main focus is on Rat advising. Every opportunity is provided to maximize and bond Rats with their academic advisor. This process begins by meeting with all Rats in each department before classes start to go over the rules and policies of the department. In addition, faculty are introduced to the new cadet class. At the end of that meeting, I reviewed the schedules of my Rats and saw if there was some way I could make a personal connection with each new cadet. At this stage, most new cadets were still shell shocked and responded with a typical "yes sir" or "no sir" response. It was also the policy of our department to allow Rats to call their parents at this time for a few minutes of "I am alive and okay" time. This act of caring was greatly appreciated by both Rats and their families.

Soon the process starts a new stage. Every two weeks until Breakout (acceptance into the Corps by upper-class cadets), we meet with our Rats for lunch in the dining hall. The Institute pays for our lunch and the whole department is there to serve as a backup in case someone is away at a conference or is unavailable for some other reason. These informal Rat lunches are an excellent way to discover cadet concerns. For example, one semester most of my Rats were failing the world civilizations (history) course. This was off the radar screen. I was used to Rats failing a foreign language, calculus, or chemistry, but not this course. A little investigation revealed there was a new faculty member teaching the course. Since I received quarterly grades for Rats (twice per semester for other cadets), I could quickly confirm a cadet's suspicions about their poor performance. So if Cadet Smith was flunking or borderline in one or more courses, I could identify him or her as an "academically at risk" student. I then invited the student to my office and discussed what he or she thought the problem was, and how it could be solved. Results of our conference were then entered into a computer database for tracking purposes.

Together we then developed an individual plan for correcting the problem. Usually my options included suggesting tutors or attendance at course reviews, seeing a counselor for personal problems affecting academic work, talking to course professors, taking papers to the Writing Center, or getting students to be more proactive in their study habits. Often the problem was poor study habits, a failure to purchase texts and the inability to manage their time in an efficient and effective manner. I am convinced the key to getting a Rat through the first year is to discover problems early and to follow-up the individual "at risk" plan, which is submitted to our departmental database.

Faculty advisors can upon request also receive information from the commandant's office, which would inform advisors if a cadet is having difficulty on the Corps side of the bigger picture. This report includes the number of demerits and penalty tours cadets have to serve. A high number of demerits or penalty tours are an indicator that a cadet likely has an attitude problem. Occasionally, students with no real direction show up at VMI. Why they come is anyone's guess, but it could be parent or sibling pressure, the opportunity to play Division I sports, the appeal of the uniform, or perhaps a warrior instinct. Again this demonstrates the importance of careful recruitment and selection of new cadets.

One year I had a football player who was a Rat. I learned early on that he was flunking all his courses except physical education. I tried several times to contact him and requested at Rat lunches that he see me ASAP; however, he never came. Into the third grading period, I learned at a Rat lunch that he had never answered any of his email from day one from anyone (cadets are required by the *Blue Book* to check their email daily). Thus at this point, his mailbox was finally full and rejecting new email messages. Now he could neither send nor receive email. Needless to say he left at the end of fall semester.

Another option I had was to suggest that students change majors, or perhaps with some Third and Second Class cadets I could suggest transfer to another major or college. While the joke on campus is that the Economics and Business Department received the engineering rejects, I think it was fair that our department populated several other majors with former students who needed a program with less emphasis on math and analytical intelligence. As an advisor, I was not there to tell cadets what to do, but to help them make the best possible decisions as they related to their VMI education and career goals. To do this I determined if they had the skills but not the motivation; had the skills and motivation, but no resources (lack of funding for tuition, books, etc.); had the motivation but lacked skills; or finally, had no motivation and lacked skills. If I could determine which profile most likely fit the cadet, then I could recommend possible solutions.

VMI Research Laboratories and Undergraduate Research Initiative

One academic area of excellence that clearly stands out at VMI is the campus wide emphasis on undergraduate research. Credit for this emphasis is owed to the VMI Research Laboratories (VMIRL). In September of 1963, the Board of Visitors established the VMIRL as the research arm of the Institute. This private, non-profit was the vision of Superintendent Brigadier General George R.E. Shell 1903, who at one time headed Marine Corps research. Shell wanted to encourage and expect VMI professors to pursue research through the administration of grants and contracts.[19] In turn, cadets would have their education enhanced by teachers who, by staying abreast in their fields, would become better teachers, and cadets would benefit by being involved in faculty-directed research projects. Over the years, VMI faculty and cadets conducted a wide variety of research projects for clients such as the National Science Foundation, National Endowment for the Humanities, U.S. Department of Agriculture, Research Corporation of America, Babcock and Wilcox, Ford Motor Company, Department of Defense, and many other state and national agencies. Recently, Colonel Barry L. Shoop, Science Advisor to the Director and Chief Scientist, Office of the Deputy Secretary of Defense, Joint IED Defeat Organization (JIEDDO) made a special presentation and plea to faculty and cadets to encourage research in the area of IED identification and counter IED measures to support the task force. Since 1963, the VMI Research Laboratories has administered more than 250 contracts and grants at a value of $10 million dollars.[20]

The VMI Research Laboratories are also noted for sponsoring the annual Environment Virginia Conference, the annual Transportation Conference, The Commonwealth of Virginia Technology Symposium (1999), and the Virginia Homeland Security Conference.

I first became familiar with VMIRL's research efforts in April 2002, when I was invited to be an external judge for VMI's first Undergraduate Research Symposium. At the time, I was employed at another college as the business school dean. The symposium is the capstone experience of the Undergraduate Research Initiative (URI). Initiated by former Dean Charles Brower, the URI is designed to more fully integrate student scholarly inquiry into the VMI experience. The program creates strong bonds between cadets and their faculty mentors. The URI is fully funded and through a series of grants available from the Wetmore Fund and Jackson-Hope Endowment, cadets can request funds for travel, software, supplies, and other appropriate expenses. In 2005, VMI and Washington & Lee co-hosted the National Conference on Undergraduate Research (NCUR). Over 2,500 students from across the U.S. flooded the two adjoining campuses.

For better students who want to pursue graduate school, the URI allows them the opportunity to develop and hone the advanced research skills sought by better graduate schools and employers. URI is also a way for cadets to tweak their intellectual curiosity and to become an expert on a topic that is of interest to the cadet and the mentor. One former cadet and Government Accountability Office (GAO) employee, Ryan Consaul (2004), wrote his mentor, "I also want you to know that the research skills you taught me when I did my independent study with you really paid off in graduate school. One of my professors was so impressed by my work this semester that he has asked me to co-write an article with him."[21]

In 2006-2007, I co-sponsored Cadet Pat McGill, who was interested in learning more about supply chain management. McGill was one of my first Rat advisees and had early on impressed me with his academic talent. Pat was one of the few in an elite group at VMI who wore academic stars for maintaining a 3.5 GPA or above. Upon graduation in 2007, McGill commissioned as an infantry officer in the Marine Corps. Realizing that I could use assistance with Pat's two-semester research project, I enlisted Lieutenant Colonel Barry Cobb, who had a background in operations management and quantitative analysis. I helped McGill develop a survey, which he sent to Virginia's top 400 employers for the purpose of collecting data. The results were later used in the development of two case studies illustrating how decision trees could be used to predict how Virginia companies would likely perform if a Pandemic flu outbreak occurred.

Once the initial research was complete, Cadet McGill presented both cases to our department and to the Big South Undergraduate Research Conference. To our delight, McGill's poster project tied for first place at the annual VMI URI presentations (a full day competitive event). Lieutenant Colonel Cobb then put McGill's research into a publishable format and submitted it to a journal on crisis management. Of course Cadet McGill was given first authorship on the submission. Another possibility would have been a submission by McGill to the peer reviewed *New Horizons,* the VMI student journal of undergraduate research or to the Economics and Business Department on-line research journal sponsored by the departmental economics honorary, Omicron Delta Epsilon. Should he later decide to attend graduate school, a published journal article would bolster his resume and application.

The Institute also offers an impressive Summer Undergraduate Research Institute (SURI) as part of its effort to enhance the academic experience and to create an environment of academic excellence. Such programs have aided the Institute in elevating academic programs to national and international standing among peer institutions. The Summer Institute once again matches a cadet with a faculty mentor to conduct research projects. This program, however, goes a step beyond and sponsors an orientation, guest speakers, social activities, and a concluding ceremony. Like all undergraduate research initiatives, its primary goal is to raise cadet standards so that cadet participation in research projects ("scholarly inquiry") becomes an integrated part of the academic experience.

Institute Honors Program

A major recruiting tool for VMI is the Institute Honors Program. Approximately 40 students are recruited annually for the program and are guaranteed admission. Others may apply if they have a minimum cumulative 3.5 GPA at VMI. To remain in the program, cadets must maintain a 3.5 average or higher and a satisfactory conduct record. The Institute Honors Program is intended to recognize a broader range of achievement than honors earned within a department. Attainment of Institute Honors is viewed as the highest academic achievement at VMI.

According to an Institute Honors brochure, "The Institute Honors Program enriches the academic experience of VMI's outstanding cadets through activities that encourage an affinity for intellectual inquiry and develop the capacity for sophisticated engagement of issues and problems, whether ethical, civic, or professional. In all of its elements, the program stresses leadership, oral and written communication skills, and the highest standards of integrity and excellence." Cadets must enroll in a Honors Forum course and two honors courses designated for sophomores and juniors. One course must be selected from the Engineering and Sciences group and one from the Liberal Arts/Leadership. In addition a senior honors thesis is required, which is supervised by a faculty mentor. The senior thesis/project ends with a public presentation. This is also true of departmental honors awardees, who present their paper to an audience of faculty and students.

Academics and Technology

Today any first class academic program must have a top priority instructional technology (IT) department. It probably goes without saying that VMI has invested heavily in technology. Through technology, VMI enables cadets to acquire superior core technological competencies; practice technological innovation, and successfully compete for the best leadership positions in the armed services, academic pursuits or employment. To accomplish this mission, VMI technology services and programs provide a secure, stable, and reliable information technology infrastructure in an efficient, cost-effective manner. It plans, implements, serves, and supports the technology needs of the Institute; and facilitates creativity in teaching, learning and communication for the VMI community through the use of competitive and emerging technologies. As a faculty member, I had all the technology needed to support my classroom activities. For example, all professors and cadets have Angel available. Angel, a widely available educational program (similar to Blackboard), enables the instructor to post assignments, lecture notes and PowerPoint presentations, video and audio clips, communicate through email, and post and calculate grades along with many other service features.

Many classrooms also use "Smart Classroom" podium equipment that allows the instructor to have one touch access to all forms of classroom technology including the Internet. Most academic buildings and classrooms are wireless; thus, I could ask my cadets to bring their laptops (99% have laptops) to class to explore web sites related to marketing examples and advertisements. To rephrase an old GM Olds commercial, "This is not your father's classroom (car)." In less than two years, I went from borrowing a laptop and lugging it to the classroom and connecting it to an Internet cable to the Smart Classroom with all its bells and whistles. Believe it or not, many faculty still prefer the blackboard ("chalk and talk" version) as cadets have become so used to technology that they often tune out PowerPoints, which can also be posted on Angel for later reviewing. Some students now call classes where PowerPoint is used "Death by PowerPoint." The old fashion chalkboard still commands attention!

To keep faculty and staff on their toes with the latest technology, it probably helps that a member of the Board of Visitors, Robert L. McDowell (1968), is currently the vice president for Business Critical Solutions for the Microsoft Corporation. McDowell leads 1,200 employees in more than 40 countries. Since 1990 he has served as the founder of Microsoft Consulting and vice president of Microsoft Enterprise Customer Services (1995-1996) and Enterprise Business Relationships (1996-2000). In 2000 he became vice president of Microsoft Worldwide Services for both consulting and product services. In his current position, McDowell focuses on identifying critical business problems that can be solved through the use of Microsoft technologies. In 2001, he co-authored HarperBusiness's *Driving Digital: Microsoft and Its Customers Speak about Thriving in the E-Business Era and In Search of Business Value*. McDowell, an economics major, got his start serving as an IT officer for seven years in the Air Force.

McDowell in an interview for the *VMI Alumni Review* stated, "I think it is important to note that, today you need basic technology acumen to be successful as a business leader. Look at the enormous impact of the Internet on business, at the enormous resources it makes available to anyone. How could someone say, 'I don't use that sort of thing?' It doesn't make sense, especially when most of your subordinates and most of your customers use it.[22]

Conclusion

Academics is the foundation of the VMI experience. Few colleges offer smaller classes or more faculty and staff to support the effort of achieving excellence in the classroom. In addition, academic advising is a major focus throughout the four-year experience, but an all out assault is made on retaining the Rat class. Rat advisors spend an enormous amount of time mentoring Rats and attempting to ensure their success in VMI's very rigorous academic curriculum. The curriculum is also enriched through the Undergraduate Research Initiative which gives cadets a chance to work with a faculty mentor to develop a research topic and to conduct research which is judged at the end of the year at the campus, regional, and national levels.

THE REGIMENTED LIFE
MILITARY TRAINING & CO-CURRICULAR ACTIVITIES

*'The Spirit of VMI, renowned in deed and
song, perhaps surpasses that of any American college.
That spirit is an intangible quality and defines
definition."*

Chauncey Durden,
Richmond Time-Dispatch, 1938

The co-curricula or military component is the fourth
ingredient in a five-dimensional framework of experiences that
helps create VMI's leaders of character. Although this book
emphasizes the academic excellence of VMI, the most distinct
and visual component of cadet life is the visitor's vision of cadets
living in an austere military environment. Whether it is the
uniforms, parades and ceremonies, or the gothic looking barracks
and old cannons, the military side of VMI permeates from all
directions. Few visitors ever venture into the classrooms, but it is
difficult to miss that VMI is a military post, not a typical college
campus.

The passage up Letcher Avenue from the Washington
& Lee campus into the VMI Post is one of stark reality. The
transformation back into time literally happens in split seconds.
Thus, this chapter delves into the daily lives of cadets and focuses
on the Corps' Regimental System that allows cadets to lead
themselves, the Class System, and ROTC or military instruction,
which prepares cadets for roles as citizen-soldiers. All three are
keys to VMI's roadmap for developing tomorrow's leaders . . . a
roadmap which leads to a road far less traveled and to the spirit
of VMI.

The old gate entrance to VMI with
Mallory Hall in the background.

Corps of Cadets Regimental System

All cadets must participate in the Institute's challenging military–based portion of the VMI experience. This includes the Corps of Cadets, the Class System, and programs associated with Army, Navy (including Marines), and Air Force ROTC. New cadets beginning the experience are called Rats until they break out of the Ratline and earn their cadet status. Once they have matriculated they sleep in the barracks, eat together in Crozet Hall, and wear the historic cadet uniforms. The process of becoming and being a Rat will be explained in Chapter 8.

The Corps of Cadets is composed of one regiment of approximately 1,400 cadets. The regiment is commanded by a regimental commander (First Captain), an executive officer and a staff of seven cadets who are responsible for all aspects of cadet daily life. In 1900-1901 George C. Marshall served as First Captain of the Corps of Cadets. The regiment is divided into two battalions with four companies each. In addition, there is a regimental band company. In total there are nine cadet companies, and 22 cadet captains.

The Corp Regimental Staff and Color Guard
led by First Captain Mathew Thompson,
march onto VMI Parade Field.
Source: VMI Communications & Marketing

Each battalion also has a commander, an executive officer
(1st lieutenant) and his/her staff (1st lieutenants) of seven who
coordinate activities with the regimental commander and his/
her staff. Within each battalion company, there is a company
commander, an executive officer, and three platoon leaders (2nd
lieutenants). All officers in the Corps are First Class seniors,
but not all First Class cadets are officers. Second Class cadets
(juniors) are selected for positions as regimental sergeant major,
battalion sergeant majors (2) and regimental sergeants (7) to

assist the regimental staff officers. Each company also is assigned a cadet first sergeant, a cadet operations sergeant, a cadet master sergeant, three cadet guide sergeants and three platoon sergeants. First Class privates carry the company guidon. Another highly prized position for Second Class cadets is to be selected as color sergeant for the regiment or battalion levels. Third Class cadets hold the rank of corporal and ten are selected per company.

The remainder of positions in the Corps are held by cadet privates, a distinctive rank achievable by all cadets that is not class based. In other words, many a senior has graduated as a private and most often they are proud of it. First Class privates also have their own organization known as the Officer of the Guard Association (OGA). According to the *New Cadet Handbook*, the purpose of the OGA is "to help create and maintain an environment that facilitates communications, harmony, and discipline in the Corps." The class system is designed to give First Class privates a great deal of power and authority. Many privates are athletes, who due to athletic duties do not have time for Corps leadership responsibilities, but they may be team leaders or club sports leaders. Also some students shy away from Corps leadership positions, if academic performance is their number one priority, either because they hope to attend graduate school or to just ensure graduation from VMI.

The ultimate responsibility for the Corps of Cadets falls on the commandant and his/her staff. The commandant is the equivalent of the dean of students at civilian colleges. A staff of three deputy commandants, three assistant commandants, two battalion advisors, the Sergeant Major to the Corps of Cadets, the Institute Chaplain, Director of the Regimental Band and Glee Club, Director of Pipe Band, and a staff of part-time tactical officers assist the commandant. The part-time tactical officers are usually ROTC staff, academic faculty, and administrative staff. They perform the duty for extra pay and are in charge of barracks in the evening (1700 to 0700) and on weekends (1700-1700). There is normally an Officer in Charge (OC) and two assistant OCs. Their responsibility is to maintain order and discipline in the barracks and to act as the commandant's representative after duty hours. They also must inspect cadet rooms three times per week on a rotational basis during the semester. In my department, several faculty members have volunteered for this duty as a way of learning more about Corps life, and as an opportunity to get to know cadets on a more personal level. It is definitely not a duty one does merely to earn the extra pay.

Cadet Daily Life

A typical day at VMI begins between 0600 and 0700 when cadets are dismissed from quarters. No VMI program activities are authorized before 0600. Once awake, cadets must quickly dress, "roll their hay" (mattress), and stack their folding wooden beds. Breakfast roll call (BRC) and breakfast begin at 0700. After a brief in-ranks inspection, cadets march to Crozet Dining Hall and are then dismissed until class. Rats are marched back by cadet cadre.

A Rat barracks room is inspected by an upper class cadet.
Source: VMI Communications & Marketing

Classes begin at 0800 and are scheduled throughout the day until 1503 on the Monday-Wednesday-Friday (M-W-F) class schedule or 1600 on the Tuesday-Thursday (T-TH) schedule. Labs are conducted in the afternoon when other classes are finished. While class schedules can vary from year to year, presently five 50-minute classes meet on M-W-F and five 75-minute classes meet on the T-Th sequence. Cadets are required to attend all classes and can be excused only with permission of the superintendent, dean of the faculty, or the Post physician. As a faculty member, I could not excuse cadets from class, although many cadets seem to think faculty can. Once the bell rings for the start of class, the section marcher (the senior cadet who used to march cadets to class) takes roll and calls the class to attention. The instructor is saluted and the instructor returns the salute and asks the class to be seated. At the end of class, the instructor is presented an absentee form, which notes who was absent and who was tardy, recorded to the minute. These forms are returned to the commandant's office, where students must answer the "Special Reports" if they are reported absent or late. This often means meeting with a member of the commandant's staff to respond. The commandant's staff determines the legitimacy of the absence and if demerits or tours are warranted.

On a side note, as an officer in the Virginia Militia and USAR (retired), I am authorized to recommend demerits for cadets. For example, I could "bone" (punish) a cadet for any infractions of the rules that I observed. The "boning" of cadets is a term uniquely used at VMI and is similar to the old Army term "gigged." When women were admitted, this term caused more than a little awkwardness, but a decision was made by former Superintendent Bunting that this traditional VMI lexicon met his standards. Other words given protected status included "Brother Rat" and "dyke."

I never "boned" (gave demerits to) a cadet, but I was very tempted on several occasions. One cadet never seemed to understand that he couldn't wear his cover (hat) in the building, even if he was on the basketball team. It seemed everyone on the faculty must have told him at least once. He delighted at being caught and always gave us a sly and knowing smile. He probably was smiling because if we did bone him in-season, he wouldn't have to walk penalty tours (PTs). Now that he is a coach, I am sure he understands the purpose of rules ... even for basketball players. Coaches generally have their own way of dealing with discipline issues; thus, it was usually best to speak directly to them. At VMI, the coaching staff is very approachable.

The biggest rule infractions observed by faculty include bringing food, snuff, or drink (other than bottled water which was finally allowed in 2007) to class; wearing an improper uniform to class such as gym attire, dirty uniforms (white shirts/pants), white socks, un-shined shoes; or having no shave or haircut. Only once have I had a disruptive cadet (talking continuously during class), and he was finally kicked out after multiple offenses across Post. VMI gives cadets every chance to succeed and to move on beyond immature, sophomoric acts; but the Institute will never tolerate honor or drugs violations. Behavior issues are normally dealt with swiftly and firmly.

While VMI is infamous among cadets and alumni for its demerit system, there is little in the way of a merit system to motivate cadet behavior in a positive and constructive manner, other than no demerits equals no penalty tours. Officials would say that good behavior is its own reward. In some ways, this reminds me of the businessperson who wants to avoid taxes. He or she can avoid them by either cheating or failing to earn a profit. The reward for earning a profit is to feel good about paying higher taxes. Right!

As the 2007-2008 chief executive editor of the school newspaper, Cadet Devin Millson stated in an editorial, "After all what is the purpose in having a demerit system if you don't have a MERIT system to counterbalance."[1] His sentiments are widely echoed within the Corps. These particular comments were made after the privilege of an optional SRC formation was taken away. This privilege allowed cadets to eat early without forming up for roll call. They could then go directly to the basketball game without returning to their rooms and then changing into their Big Red spirit t-shirts. Cadets feel these small privileges, especially their treasured class privileges, help provide them small amounts of free time to catch up on sleep or just take some time off from a relentless schedule. These types of cadet issues come and go depending on the commandant and superintendent. Certainly, rewards for positive behavior are within the realm of motivational theory and practice and are common at other military institutions in the form of weekend passes and other small breaks from routine life.

On Mondays and Wednesdays (spring 2009), the third class period from 1100 to 1215 is reserved and shared as either Commandant's time or Dean's time. The Dean's hour is often used for guest speakers, advising, departmental class meetings, common exams, convocations, make-up classes and tests, and other times so designated and reserved by the dean, department chair or faculty member. The Commandant's time is used for a variety of purposes ranging from inspections, parade practice, professional development training, military instruction, and human resource issues.

Military Duty and Physical Training

Once classes are over until Supper Roll Call, cadets are free to engage in a variety of activities ranging from drill and ceremonies, intercollegiate and club sports, or physical training (PT). PT is mandatory every Monday from 1615 to 1750 and Thursday from 1615 to 1750. ROTC programs have the primary responsibility for physical training. Any unscheduled time reverts to cadet time. For example, if a Friday afternoon parade is cancelled because it is raining, cadets are then free to reschedule the time to their benefit. Practice for parades is normally held on Thursday afternoons at 1445.

Probably the most common use of afternoon time is the personal workout (PWs) and/or jogging time, which is no longer mandatory effective 2008-2009. It is a very common sight in Lexington to see cadets jogging through town, individually or in groups. Every semester ROTC units do group runs through town led by the professor of military, naval/marine or air science and his/her cadre. I also notice a high number of male cadets running through the W&L campus, either eating their hearts out or hoping to be noticed by some young "mink" beauty. With a goal of always keeping the body fit, a run through the W&L campus sorority row could easily be the highlight of their day. One thing for sure, it beats walking penalty tours (PTs) on Wednesdays from 1615-1550.

Other cadets may choose to engage in club sports, which are very popular at VMI. Teams usually practice from 1615-Supper Roll Call (SRC) hours. There are two SRCs. Beginning with the 2008-2009 academic year, the designated battalion marches to the dining hall at 1810 and the remaining battalion and other in-season athletes at 1900. Late supper for in-season athletes (football and basketball) is held at 1920. Teams include fencing, boxing, rugby, marathon/triathlon, ice hockey, golf, cheerleading, sprint football, martial arts, men's soccer, and women's softball. Rats may try out and participate on a team, but once selected are normally only allowed to play at an apprentice level. Their participation in the fall is limited to Wednesdays due to Rat Challenge on Tuesdays and Thursdays. Thus, they usually cannot compete in actual games during their Rat year. On Fridays, cadets participate in a dress parade from 1630-1730. Finally, the afternoon ends with SRC formation and the march to the dining hall.

Evening Study Period

Evenings from 1930-2315, Sunday through Thursday and Fridays 1930-2345 are reserved for study, homework, and course preparation. No non-academic activities may be planned for this time so that cadets can focus on academic excellence. Cadets who have chain of command responsibilities or class leadership duties may conduct them until 1930. Individual officer, sergeant, and class duties may be performed until 2030. Cadet corporals and privates may not perform any chain of command or class duties during these hours. They may volunteer for tasks if they are authorized according to the privileges of their class and rank.

Cadets are also authorized to schedule one extra-curricular activity from 1945-2100 for a maximum of 75 minutes on Mondays (clubs) and Tuesdays (religious meetings and mass activities). Tuesdays are also reserved for Honor Court assemblies and class meetings. Mandatory lectures (Dean's Lecture period) are conducted on Wednesdays and Thursdays.

From 2315(CQRB) to 2330, cadets must return to their rooms unless otherwise authorized to stay out. Taps play at 2330 Sunday-Thursday, 2400 on Fridays, and 0100 on Saturdays. Once taps has blown, desk lamps and computers are authorized until 0130 for First Class cadets and Second Class cadets, 0030 for Third Class cadets and until Taps for Fourth Class cadets. Extended late study hours may be requested, but all cadets should be in their rooms with lights out and asleep after 0130. Based on the time that many emails were sent to me and cadet confessions, I have serious doubts if cadets closely follow the order of going to bed at the required times, or there are lots of official exceptions made.

Guard Team

One of my least favorite memories as a cadet or an officer was guard duty. For one thing, it meant we had to stand inspections, which were known to produce more than their share of demerits and stress. Second, who likes to stand around in the hot, cold, rain, or snow? When I pulled Officer of the Day (OD) duty in Vietnam, I was never sure when our guards might shoot me. Soldiers frequently slept while pulling guard in our perimeter towers. It was my misfortune, but duty to climb the tower steps and to ensure that they were alert. After the first time of nearly getting my head shot off, I always called out if I was not challenged; if then there was no response, I threw a stone into the tower before climbing the steps through a closed hatch in the floor. There was also the chance that if they weren't sleeping, they were smoking pot or heroin. Trying to surprise them was not a good choice in an environment of frequent "fraggings" and shooting of officers by our own troops. My battalion had two fraggings, which injured nine soldiers. In 1970-1971, the years I was in Vietnam, the Army reported 363 actual fraggings in Vietnam and an additional 118 possible fraggings. The result of these senseless acts was 45 killed and countless numbers wounded. For the same period (1970), the Army reported 14,571 drug abuse cases.[2] These incidences illustrate a failure of leadership at all levels of command.

At VMI nothing quite so traumatic occurs, but 24 hour guard duty is an essential safety feature for the Corps and a time-honored tradition of ceremony. After the Virginia Tech massacre, I was asked by a parent at an open house for prospective cadets about security issues at VMI. My response was it's hard to beat 24 hour security by armed (but not loaded and absent of firing pens) guards posted at every entrance to the living areas. Post security is also supported by a full-time and fully-armed Post police force. The guard team also enforces the rules and regulations of the Institute. Further, it acts and reports all emergencies, to include fires, water main breaks, medical emergencies, and others. Some guards (usually Second Class cadets) act as orderlies and assist visitors at information booths in Jackson Hall and check IDs in Cocke Hall for those using the gym facilities.

The guard is manned by Fourth Class cadets, who march one of three posts as sentinels or serve as supernumeraries (pages). Cadets in the upper classes command the guard team in different shifts, which are organized by the company operations sergeants. The cadet OD (Officer of the Day) and cadet OG (Officer of the Guard) can be easily identified by their maroon sashes, sabers, and keys. Traditionally the OD dangles a mass of keys, which make a great deal of noise to warn cadets as they come through the barracks. The guard team cadre is quartered in the Old Barracks guard room inside the entry of Jackson Arch and adjacent to the commandant's office. From there they make the many PA announcements during the day and evening. These include warnings before formations and all other relevant information needed by cadets for the day including which uniform to wear.

Cadet guards post at Washington Arch
at the "Old Barracks."

Guard duty is an excused absence for cadets from class attendance. Hopefully, no cadet will pull guard more than two times per semester, and thus it shouldn't affect class performance. On occasion, I've seen cadets miss class twice in two-three weeks due to guard, but guard normally follows a regular rotation schedule for Rats. Every cadet company pulls guard every nine days, but not everyone pulls guard on every rotation. Upper-class cadets can sign up for when their schedules permit. At this point, I usually follow-up with the cadet because faculty do not receive advance notice that a cadet will be on guard.

Weekends

If you think weekends are probably a piece of cake, think again. While there are rarely activities before 0800 on Saturdays, in the early fall Rats occasionally train with cadre by doing road marches in preparation for the New Market march. There may also be a few isolated re-inspections of delinquent rooms. Generally, cadets can expect a full morning of activities in the fall including morning parades followed by a home football game (usually six).

Parades at VMI are special and VMI cadets are capable of putting on a show that is equal to any service academy. With House Mountain looming in the background, the band plays military marches and the pipe band in kilts adds an extra flair to the pageantry. The band director, Colonel John Brodie, is a former enlisted Marine and the honor man of his boot camp. The boyish looking Brodie (over 20 years at VMI) is adored by Corps members and seen at nearly every major event from parades, hockey games (former player/coach), and athletic events. Brodie is credited with building the band to a unit of over 115 members, the largest size since its founding in 1948. Brodie prides himself in "making a lot of rock guitarists into trombone or trumpet players."[3] He is an honorary member of the class of 1992 and was named the 25th honorary alumnus in 2008.

Band Company represents VMI in many
state and national events.

Saturdays are divided into three categories from 0800-1200. These are designated annually by the superintendent. Academic duty Saturdays are reserved for academic and academic support departments. Once reserved, other departments cannot schedule this time without permission from the dean. This time may be used for mandatory or voluntary activities. All unscheduled time reverts to cadet time for personal use. Military duty Saturdays are reserved for the commandant. In the fall these may be used for parades before each home football game. ROTC Duty Saturdays are reserved for use by the ROTC departments. These weekends are used frequently by the ROTC units.

On Sundays cadets are released from quarters at 0830 and may attend an optional brunch served from 0830-1300 hours. Many cadets will go into town and attend religious services (especially Rats). No mandatory activities may be scheduled prior to 1300. Penalty tours (PTs) may be marched between 1300 and 1600. Generally, Sundays are a good R&R day, or as many a cadet has done, cruise the Valley visiting Mary Baldwin, Sweet Briar, and Hollins colleges and coed universities like James Madison or Virginia Tech.

Two weekends a year are devoted to a field training exercise (FTX). The spring FTX extends through Tuesday SRC; and thus, I had the good fortune to participate in nearly every FTX weekend since my first year at VMI. For the past three years, the VMI Air Force ROTC (AFROTC) detachment has traveled to our outdoor education center to conduct one phase of their FTX. On weekends, I manage a non-profit leadership training organization I founded called Goose Creek Adventure Learning, Inc. Since its conception in 2002, Goose Creek has worked with over 1,200 youth and young adults from across the nation at our 52-acre facility in Bedford County near Smith Mountain Lake. Our board and leaders are unpaid volunteers. Goose Creek's main attraction is a leadership reaction course (LRC), which is patterned after those found on military bases. The corporate world has also discovered the value of such courses, which teach leadership principles through physical and mental challenges. For example, each year the incoming class of MBAs at Wharton has the option of participating at the Marine LRC at Quantico, Virginia. In the near future, VMI will have its own LRC as part of the new $15 million Leadership Field Training project.

These courses are excellent exercises in problem solving, decision making under pressure, teamwork, communications, physical fitness, creativity, resource management, time management, and other core leadership elements. Cadets, utilizing a different leader each time, rotate between 10-12 different stations, each providing a unique physical and mental challenge. When working with the AFROTC, advanced cadets debrief and evaluate each team leader and team for their performance at each station. In addition for AFROTC cadets, we run a "downed pilot" exercise, which requires cadets to use a GPS to navigate around the 52-acre terrain seeking clues to the whereabouts of the pilot. I also use cadets from my Business Leadership course to lead and evaluate JROTC cadets from regional high schools. They attempt the course as an activity sponsored by the Piedmont chapter of the Military Order of World Wars (MOWW).

Using VMI cadets to run the LCR for Junior ROTC (JROTC) cadets has proven to be an important element to capping the VMI experience and to ensure that more cadets are given leadership opportunities. The value of the LRC is summarized in this comment from a cadet leadership journal, "I went into that Saturday at Goose Creek thinking that I was going to embarrass myself and it was going to be awkward being with those students [other cadets in class] who are better at leading than I am. I had an extremely low opinion of myself and was waiting to fail. The results of that day, however, were a big turnaround for me. The self esteem boost I got from knowing that I hung in there with others really motivated me for the rest of my college experience and more likely for the rest of my life. A little bit of responsibility changed the way I thought of just about everything."

Two colleagues and I worked with the Air Force ROTC detachment leadership to evaluate the validity and reliability of LRC cadet leader evaluations. We tracked cadets as they participated in the Air Force summer field training program and evaluated whether the Goose Creek LRC results were accurate

VMI Air Force ROTC cadets problem solve at
the Goose Creek Leadership Reaction Course (LRC).
The LCR is an important part of Air Force ROTC cadets' preparation
for summer training at Maxwell Air Force Base.

predictors of performance at field training. Obviously, this research project could also be partnered with a cadet as an Undergraduate Research Initiative project. In the summer 2007 field training, the performance of VMI cadets was clearly superior to their peers from other colleges. Approximately 30 percent of the Air force cadets finished in the top 20 percent. Eight cadets finished in the top 10 percent and earned distinguished graduate status.

In 2007, it was my good fortunate and privilege to work with Cadet George "Will" Flathers III on this project as a part of his Air Force leadership lab. Cadet Flathers, an electrical engineering major from Elkton, Virginia, was selected in 2007 as VMI's first Goldwater Scholar and then late in the year was named VMI's first Marshall Scholar. As a Marshall Scholar, Flathers is spending two years at the University of Sheffield in England. He is studying Automatic Control and Systems Engineering at Sheffield. Flathers hopes to work someday at NASA's Langley Research Center. From my perspective, he was a "squared away" cadet.

Flathers was the Regimental S-2 in 2008 and an Honors Program student. Lieutenant Colonel James Squire, Flathers' academic advisor has said, "Will is a very caring individual and that concern for others drives his research interests." Flathers and Squires joined forces in 2006 to pursue "a way to experientially connect parents with their hearing impaired children." His software program, which he developed, "simulates a patient's hearing to give parents, physicians, and educators a better understanding of the sounds the children hear."[4] Brigadier General Brower supports Squire's observations. "What impresses me most about Will is that he is not narrow in his focus but has an agile, restless intellect; he questions and thirsts for understanding. His intellect combined with inner strength and determination is convicting, and I'm sure he will build upon this opportunity in what promises to be a very successful future."[5] Flathers commissioned in the Air Force upon his graduation in 2008.

Extra-Curricular Activities and Clubs

Like most colleges, VMI has a wide variety of extra-curricular associations. Anytime there are a few cadets interested in an activity the potential for a club exists. The naturals are the Institute and departmental honoraries and academic discipline clubs. Most clubs are happy to have new members, but a few are selective. For example The Cadet Investment Group is open to all majors on campus; however, its two advisors, currently Colonels Cliff West and Bob Moreschi, traditionally have come from the originating department of Economics and Business. The advisors and cadets work closely with VMI Foundation, which oversees the fund. The Cadet Investment Group is one of the most astounding activities available to any student, anywhere. Selected cadets are divided into two investment teams. Each team elects its own officers and invests a real $100,000 cash portfolio (total of $200,000 available for the club). Members are assigned a stock of their choice to research and manage, and members decide to buy or sell as the year progresses. During the year, the club takes a trip to Wall Street in New York and/or visit the Federal Reserve in

Richmond. At the end of the year, stocks are sold and an awards banquet is held when new officers are announced for the next year. Any profits earned are plowed back into the endowment fund for next year's members to invest.

Other categories of extra-curricular activities include VMI publications (*The Bomb* yearbook and *The Cadet* newspaper); religious clubs (Newman Club, Officers Christian Fellowship, Outreach Club, etc.); athletics (Cheerleading, Racquetball Club, Riding Club, Rugby, Power Lifting Club, Road Runners, etc.); Political (College Democrats, College Republicans, and Pre-Law Clubs); Military (Army Aviation Club, Ranger Challenge), Pistol and Shotgun Clubs, Volleyball Club, etc; Health (EMT/ Firefighters and Pre-Allied Health) and Music and Theater Clubs (Glee Club, etc.). These clubs are but a few of the many extra-curricular activities authorized for cadet participation and leadership opportunities. It is not unusual to find a cadet private running a club sport. For example in 2007-2008, Cadet Drake Watts, a First Class private was president of the Rugby Club as was the present commandant Colonel Thomas Trumps (1979). This leadership role takes a tremendous amount of time commitment and dedication, which includes scheduling games, coaching, managing club resources (dollars and equipment), arranging transportation and many other duties.

The Cadet Investment Club, which annually manages a
$200,000 portfolio, poses on Wall Street.
Source: Col. Bob Moreschi

Cadet Drake Watts (2008) was president of the Rugby Club and a member of Band Company and the Drum and Pipe unit.

Military Training and Instruction:
Every Cadet A Soldier

Few Americans know what it truly means to serve in the armed services of our nation, especially during a war. Most can only guess and attempt to satisfy their lack of understanding by playing a Combat video game or paintball. In October of 2005, I was sitting in Boston's Logan airport awaiting my flight and reading *Jar Head,* when I noticed people moving past me towards the large side window. A crowd starting gathering and I was drawn to get up and see what was happening. On first glance I saw a commercial aircraft, which was surrounded by fire trucks, state police cars, Marines, and others. My first reaction was that a dignitary was arriving or perhaps the aircraft had experienced an emergency. Soon a black limousine and hearse drove up to the plane and a family stepped out with a military escort. A sudden and indescribable feeling washed over me as I soon watched a flag-covered casket slide down the conveyor belt towards a waiting Marine Honor Guard. The Guard stood at rigid attention as the family gathered around the casket and embraced one another. It quickly sunk in that I was watching a scene that most Americans will never witness ... the return of a dead Marine or soldier to his or her family.

Of the hundred or so people standing at the window, there was hardly a dry eye. A few vets in the waiting area stood at attention and saluted as the Honor Guard took hold of the casket. There was nothing in that busy airport corridor but dead silence. Time seemed to stand still and for that moment perhaps it did. The war with all its horrors had finally come home to America. All those watching were thinking that could be them in that casket or it could them collecting their child from strangers in a distant place. We all grieved for the family of that lost Marine.

The image of that scene is locked in my mind. It is a constant reminder that every American needs to understand the tremendous sacrifices that these brave soldiers, sailors, airmen, and Marines and their families are making each and every day. Our nation is truly blessed that VMI trains and produces young people, who are willing to lead our troops and nation with integrity, courage, and selfless service.

Since 9/11 and General Peay's leadership on Post, there has been a big push to increase the number of commissioning cadets. Prior to 1988, all cadets had to accept a commission as an officer if offered one. This practice ceased about the time of the Soviet Union's collapse and the perceived U.S. need for a smaller military. At the time of 9/11, the percentage of cadets commissioning had dropped to approximately 34 percent. Using his connections and the strong need for junior commissioned officers, General Peay was able to obtain more ROTC scholarships for VMI cadets. By 2008, 51 percent of the graduating class accepted a commission in the Army, Navy, Marines, or Air Force. Of that 51 percent, approximately 18 percent will make the military a career. By November 2008, 1,200 VMI alumni had served in Iraq and Afghanistan.

One of my favorite movies growing up was *The Long Grey Line*. I remember a scene where graduating cadets dropped by the home of West Point's long-time boxing instructor Marty

Maher. Over time, one by one the cadets said goodbye to Maher and reminisced about friends who had gone before them to the battlefields of World War II. When I started teaching at VMI, I quickly realized that something different was happening at the Institute. In my first month of classes in 2002, I had a student tell me he would be dropping my course, because he was deploying with his Army National Guard unit. I had never given this much thought until one by one over the next six years I would see at least four of my students deploy. By the end of 2008, 63 cadets and seven faculty and staff from the Institute had deployed to fight the Global War on Terrorism.

In addition, several faculty and staff have deployed to bases around the world. One faculty member and former department head from the History Department, Colonel N. Turk McCleskey, USMCR, was called out of retirement to serve on an administrative board at Guantanamo Bay, Cuba for the purpose of reviewing prisoner records and determining their status for possible trial or return to their native countries. This was a tough and controversial job, which Colonel McCleskey performed with a high degree of proficiency and sense of duty. He was often quoted in the news as an unidentified Marine Colonel of authority. His role in the Global War on Terrorism was one of great importance as the U.S. struggles to handle "enemy combatants" and terrorists, who are not traditional soldiers from state nations recognized by the Geneva Convention. His leadership switch from department chairman and faculty member to military duty performed in adverse conditions provides a reminder of the Old Corps faculty. Many left VMI to command Southern army units during the Civil War. Once again the VMI faculty has set the example for its cadets.

Cadets from Guard and Marine Reserve units in Lexington, Roanoke, Staunton, Lynchburg, and Harrisonburg have sent VMI cadets to such places as Iraq, Kosovo, Cuba, Afghanistan, and Qatar. What strikes me is the immediate sacrifice these cadets make to serve their nation. If they return (most do) to VMI, they will likely have lost a minimum of 18 months of their life and education. They no longer have the opportunity to graduate with their Brother Rats. For example, Mark Miller (2007) was wounded in an ambush on his first Iraq tour with his Marine reserve unit and recently returned from a second tour. When talking to a "Welcome Back Warriors" reception, Miller showed a news clip of a firefight in Iraq that took the lives of four members of his Virginia unit. Another cadet, Ryan Koniak (2005) missed one academic year when he was called to serve with his Army National Guard unit in Guantanamo, Cuba. Koniak's outstanding performance resulted in his being named the Virginia's National Guard's Soldier of the Year. Representing First Army East, he finished five points from winning the national award.

Jonathan Faff (2009), a sergeant in the USMCR was in my Principles of Marketing course. He had already deployed three times to Iraq and Afghanistan and had more ribbons than most senior NCOs and officers prior to the war. In talking with Jon, I could tell his mind was occasionally somewhere else. I did not want to push him to talk about his experiences during the invasion, as I knew his unit was in the thick of things and took heavy casualties. He shared that he felt most comfortable when he was talking with other veterans at the local VFW. I understood that his brothers were his fellow Marines. Early in 2008, I learned Jon had volunteered again and was deployed for a fourth tour in Iraq. He is now back in classes, and to date, it is my understanding that he is VMI's most decorated cadet. No doubt Sergeant Faff has seen more combat than most Marines assigned to active duty units. Semper Fi!

In late 2007, Cadet Jake Jackson (2010), returned from Kosovo after a 15-month deployment and once again enrolled in my marketing course. Jackson also has four rows of ribbons and a positive and outgoing attitude to go with them. In Kosovo, Jackson served as the bravo team leader in his squad and was responsible for training his team in common task skills and conducting patrols. The mounted team used Hummers to patrol the countryside in rural areas and dismounted for village patrols. In villages, they were often tasked to interview villagers using native translators. In February of 2008, Jake was just settling back into the VMI academic routine, when he and 10 other cadets were called back to active duty by the governor to fight forest fires in Central Virginia. In September of 2008, he once again activated to standby for Tropical Storm Fay. For his service, Jackson had been awarded three Army Achievement Awards and received the Army Good Conduct Medal.

Another cadet in Jackson's marketing section, Benjamin "Finney" Kimsey (2009), a psychology major, didn't need to wear his ribbons. Kimsey wears the coveted Combat Infantry and Paratrooper Badges. Before graduating from high school in Harrisonburg and two weeks after his 17th birthday, Kimsey enlisted in the local 29th Division Army National Guard unit. While enrolled at Virginia Commonwealth University, he was deployed to Cuba, where he served nine months performing a variety of base security roles. After returning home, Finney discovered that his twin brother, David Kimsey (2007), also in the Army Guard, was being deployed (13 months) to Afghanistan. Kimsey then volunteered to deploy with his brother and the two served together at a forward operating base south of Kabul. Nineteen days after returning home, both brothers were enrolled at VMI.

David Kimsey, having a head start on his brother Finney, was able to graduate one class behind his BRs of 2006. He is now an infantry officer (Ranger) serving with the 82nd Airborne. Finney wasted no time as a Rat and was elected 2009 class president. Through observation, I have been most impressed with Kimsey's strong sense of professionalism as he enforces the class system, a responsibility, which as class president he takes very seriously. Telling your peers to shape up isn't easy, but when they respect you as they do Kimsey, it is much easier. After graduation, he will also commission and serve additional time in the Virginia Army Guard as an officer. Somehow both Jackson and Kimsey at my request found the time to drive to Lynchburg to address a Military Order of World Wars (MOWW) dinner meeting.

Jon Glasscock (2006) returned to VMI just 45 days after returning from a deployment in Afghanistan. He served as a medic in a 12-person field hospital located in a desert in the

"Finney" Kimsey, class president of 2009 and an
Afghanistan War veteran, is responsible for
ensuring the class system is adhered to by all cadets.
Source: VMI Communications & Marketing

Ghazni Province. Glasscock said of his experience, "It's been a tough road and a long five semesters. Still, I'm planning to take my experience and put it to use as a Virginia Guard officer and head back over as soon as I can."[6]

Another guardsman, Cadet Sarah McIntosh, was a 20-year-old Third Class cadet, when she deployed to Iraq. McIntosh deployed in the summer of 2007 as a member of the Virginia Army National Guard 429th Brigade Support Battalion Transportation Company, which she joined as a junior in high school. While other female VMI graduates have deployed to Iraq, she was the first female member of the Corps to deploy to Iraq. Her duties as a truck driver involved dangerous convoy duty hauling supplies from Kuwait to Iraq. McIntosh desires a career in the Army once she graduates. McIntosh decided when she was six years old that she wanted to attend VMI. Both her parents are veterans and her brother Joe was also a First Class cadet at VMI.[7] Sarah feels VMI reinforced her military skills and made her better prepared for her hazardous duties. Another female, Emily Fritts, deployed in 2008. On a personal note, it was a privilege and an honor to have taught many of these young men and women.

Sarah McIntosh (2010) with members of
her Army Nation Guard unit in Iraq.
Credit: Courtesy of Sarah McIntosh

As an additional element of VMI uniqueness, General Peay worked very hard at the state and national level to institute an actual Virginia Army National Guard detachment at VMI beginning in 2008-2009. The purpose of this unit is to further strengthen the concept of the citizen-soldier. The Guard each year awards approximately 40 full-ride VMI academic scholarships to cadets willing to serve and train in the Virginia National Guard and be commissioned as second lieutenants upon graduation. This novel idea received support at higher levels and will help the Guard and other military units fill badly needed junior officer slots. It will also make VMI the first and only college or university with its own Guard unit, and certainly help the Institute fulfill its mission of producing citizen-soldiers.

Reserve Officer Training Corps

The backbone of all military instruction occurs in the Army, Navy/Marine, and Air Force ROTC departments. The mission of each ROTC program supports the mission of VMI. The Institute has three of the finest and largest ROTC departments in the nation, each with a long history of service to the Post. Kilbourne Hall, the newly rebuilt home (2008) of the ROTC programs is said to be the largest and most modern building in the U.S. dedicated strictly to ROTC programs.

The seed of Army ROTC was first planted in 1913 by a committee of college presidents from Harvard, Princeton, Yale, Michigan, and Alabama as well as VMI's superintendent General Nichols. With General Leonard Wood's leadership, they organized the Society of the National Reserve Corps. ROTC became a reality in 1916 with the passage of the National Defense Act. In the same year, General Nichols announced that the Institute would establish infantry, cavalry and artillery units. The birth of these units had to wait until the end of World War I. In 1924, an engineering unit was also created. The first group of cadets commissioned in 1921 as reserve officers.

During World War I and before the Students Army Training Corps (SATC), VMI ran three camps for training civilians. A faculty member, Captain Henley P. Boykin, served as camp commandant; and the superintendent was commissioned as an army major and made director of the SATC program. First Classmen served as instructors for the camps. Dress parades were an interesting sight during this period as VMI had not only an army unit training, but also one of 15 Marine training units. Thus, there was a mélange of uniforms in cadet gray, olive green, and Marine blue. Of course the Marines stood in the front row.[8] Air Force ROTC was instituted in 1946 and Naval ROTC in 1974.

Army ROTC

According to the U.S. Army Cadet Command, the mission of Army ROTC is to commission future officers and to motivate young people to be better citizens. Cadet Command establishes partnerships with American universities and provides them an "opportunity to better understand and shape the values of those who choose to follow the Profession of Arms." In turn, colleges and universities gain talented men and women. More than 40 percent of the General Officers in the active component of the Army receive their start through ROTC, and 60 percent (18 percent West Point) of newly commissioned lieutenants enter active duty through the Army ROTC program.

VMI's Marshall-New Market Battalion like other VMI ROTC detachments is composed of cadets from Mary Baldwin, Southern Virginia University, Washington & Lee University as well as VMI. The focus of the program is leadership development in both classroom and field environments. In addition, the battalion offers a Cadet Battery and a Ranger Platoon. In 2007, 66 scholarships were awarded to incoming Rats. The unit goal is have 80 percent of Army cadets on scholarship.

The first commanding general (CG) of the U.S. Army Cadet Command (1986) was Major General Robert E. Wagner, VMI class of 1957. While in Vietnam, Wagner served three years in direct contact with the enemy. "His heroism under fire resulted in five separate decorations for gallantry, including the Silver Star and multiple Bronze Stars with "V" device" and the Combat Infantryman Badge.[9]

Major General Wagner is credited with putting his personal stamp on all aspects of cadet training, especially with innovations in training and development, leadership evaluation techniques, and his toughening of training standards. Cadets enrolled in Army ROTC take several courses in leadership, teamwork, ethics, problem solving, management, and the capstone officership course. Each cadet will also complete a lab and a field training exercise (FTX) each semester. Although retired, Wagner remains active promoting the Army JROTC program (300,000 strong) and spearheads the annual Army ROTC Leadership Symposium partnered with the George C. Marshall Foundation.

Between the summer of the Second and First class years, all commissioning cadets complete an eight-week CFT (cadet field training) at Ft. Lewis or other designated posts such as Ft. Bragg or Camp Buckner (United States Military Academy at West Point, New York). One of the eight weeks will be spent at Ft. Knox, Kentucky, where cadets will complete Mounted Maneuver Training (MMT). To graduate successfully cadets must pass land navigation, the Army Personal Fitness test (APFT), Combat Fitness Test (CFT PCT-1) tasks, water obstacle course, maneuver light training, and qualify with the M-16. Scholarship cadets also have the opportunity to attend Airborne School, Air Assault School and Combat Diver Qualification (4-week SCUBA school). Most Army ROTC cadets are on a three- or four-year scholarship. They earn the same monthly stipend ($250-$400), tuition and fees, book allowances, and lab fees as cadets in other ROTC programs.

Air Force ROTC

The Aerospace Studies Department's mission is to produce commissioned leaders for the Air Force and like the Army to build better citizens. Program graduates can choose from over 45 career fields open to officers including pilots, navigators, weather officers, security officers, and combat rescue officers. Cadets must first enroll in a two-year General Military course, which consists of one hour of classroom instruction and two hours of leadership laboratory each week.

Cadets then must be selected for the remaining Professional Officer course. They are then enlisted in the Air Force Reserve and assigned to the Obligated Reserve Section. Selection into the program depends on GPA (2.0 required and C or better in AFROTC courses), and medical and physical fitness results.

The Professional Officers course requires three days a week of classes in Air Force Leadership and Management, National

Security Affairs, and a once per week leadership lab which is led by advanced cadets. Rising sophomores are eligible to participate in a three-week summer program to shadow Air Force officers, tour bases, and get some hands-on experiences in a unit. Cadets spend the summer between the sophomore and junior year at various air force installations such as Maxwell Air Force Base in Montgomery, Alabama, where they complete a mandatory 29-day field training. In field training cadets practice marksmanship, a confidence course, drill and ceremonies, classroom instruction in leadership and human relations, survival training, and participation in Air Force specialty and aircraft and crew orientations.

Other opportunities abound for scholarship cadets and include but are not limited to a 15-day course in soaring (non-power) at the Air Force Academy, internships at various Air Force units, Army Airborne School (three weeks), and a NASA program.

VMI cadets, who desire to fly by far exceed the national selection rate of 55 percent. In 2004, of the 15 cadets choosing to pursue a flying career, 14 were selected (10 as pilots and four as navigators) for an overall rate of 93 percent.[10] One former cadet, now Lieutenant General Daniel J. Darnell (1975), served as commander and leader of the U.S. Air Force Thunderbirds flight demonstration team. Darnell currently serves as Deputy Chief of Staff for the Air, Space, and Information HQ, U.S. Air Force, Washington D.C.

As a side note, other VMI aviation alumni seem to think they are also members of the Thunderbirds or Blue Angels. Anyone living in Lexington eventually gets used to the regular low altitude flyovers and occasional sonic booms. Of course, it is also possible that flyovers are practice bombing runs by former cadets, who are still dealing with their four years of endless frustration.

Navy/Marine ROTC

According to the VMI NROTC web page, the mission of the Naval ROTC unit is to "develop midshipmen mentally, morally, and physically and to imbue them with the highest ideals of duty, honor, and loyalty in order to commission them as naval officers who possess a basic professional background, are motivated towards careers in naval service and have a potential for future development in mind and character and are prepared to assume the highest responsibilities of command, citizenship, and government." Navy option midshipmen are commissioned as unrestricted line officers. They can request duty as naval aviators, SEALS, submarine officers, and surface warfare officers among other options. Marine officers can serve in either ground or aviation units. In 2007, VMI NROTC was the only NROTC unit in the nation to graduate 100 percent of its cadets/midshipmen.

While Navy midshipmen spend their summers on cruises assigned to naval vessels, Marine cadets attend Marine Officer Candidate School (OCS). In 2007, Robert Jordan (2007) achieved the highest leadership average (99.2) of 600 graduates in Marine OCS. Elizabeth Dobbus (2009) was awarded the Platoon Leaders Class "Gung Ho" distinction of 300 candidates.

Pilots have a choice of fixed wing/attack aircraft, rotary (helicopters), or multi-engine transportation aircraft. During the initial invasion of Iraq, four VMI graduates commanded U.S. ships, and two commanded one of the Navy's 13 aircraft carriers. Aircraft carriers were commanded by Captain Thomas Parker (1974) (USS Kitty Hawk) and Captain Charles Smith, class of 1979 (USS Eisenhower).

In 2003, Marine Major Daniel Shipley (1992) served as a member of the prestigious U.S. Navy Blue Angels flight demonstration team. Prior to his selection, he served seven months aboard the U.S. Constellation flying combat operations over Iraq. As a VMI cadet, Shipley was cadet lieutenant and battalion S-5. He was a member of the lacrosse team, the

Religious Affairs Committee, and Officer's Christian Fellowship. In addition, Major Shipley was also a member of the Honor Court during his First Class year.

Standards for acceptance in the NROTC program are high. A 2.5 average is required and unlike other ROTC programs, midshipmen must take calculus or college math, physics, a physical science, a computer science course, and others from the VMI curriculum to complete the NROTC program. NROTC courses include introduction to Naval Science, Sea Power and Maritime Affairs, Leadership and Management, Navigation, Amphibious Warfare, Naval Engineering, Naval Weapons Systems, Evolution of Warfare, Ethics, and either Navy or Marine leadership labs. Navy scholarship midshipmen are encouraged to declare a major in engineering or a hard science to meet the technical demands of the Navy.

The Importance of Mission:
Planning and Execution

My work at the Goose Creek LRC with the VMI and University of Virginia's AFROTC detachments gives me a closer look at how the Air Force trains its future leaders. For sure, one thing really impresses me about Air Force training: The Air Force cadre work hard to instill in cadets the importance of planning meticulously down to the last detail. This is drilled into them on every occasion possible. Cadets in leadership roles are held responsible and accountable for their training results. It pays to get it right by using a proper planning process based on a concise mission and objectives.

Mission Statement and Objectives

For the first training mission to Goose Creek, I was invited to a pre-briefing, where the Cadet Officer in Charge (COIC) immediately provided me a handout with the mission for the activity stated in a clear and understandable manner. The cadet mission was "to transport the 57 AFROTC cadets to Moneta,

Virginia by 0830 for summer pre-training to train advanced cadets how to properly evaluate and debrief participants. Cadets are to return no later that 1700." Each part of this mission objective was measurable and realistic.

Since the mission statement said the training would take place in Moneta, Virginia, then one would expect that everyone driving would know how to get there. Verbal directions to the property were reviewed with cadet leaders who said they understood. Understanding the conditions of a mission statement is very important. For example, if cadets showed up at the wrong location, the mission could fail. After a long wait on our first training mission, I learned the cadets were lost. I quickly discovered that ROTC cadets also make mistakes. That is why they are cadets. When the first group of cadets arrived over 40 minutes late, cadet leaders had the responsibility of adjusting the schedule to optimize their remaining time. The next time they went on a training mission to Goose Creek, everyone had maps and leaders carried cell phones to communicate between buses and vans.

Identify Threats

If you are a business manager, a business student, or a veteran or member of another branch of the armed services, undoubtedly you recognize that the Air Force plans its missions in a similar manner as every business or organization should. While threats in business could be your competitors or conditions in the environment such as the economy or federal laws and regulations, military threats are usually potential enemies. In the case of this training exercise, cadets needed to think about the weather as a threat or hidden threats such as a vehicle that might break down or a serious injury to a cadet, which might require emergency assistance.

Identify Support Assets

Cadets quickly learn they need to communicate with the officer cadre to coordinate and ensure they have the support needed to complete their mission. To prepare for the mission and possible threats, our cadets read local weather forecasts and inform participating cadets to bring an extra change of clothes if needed in case of rain. The Field Training Exercise (FTX) continues in all weather conditions with the exception of lightening or extreme heat. Cadet leaders also ensure that all on the team bring their own water, MREs (food), and a first responder or EMT in case of a medical emergency. Another issue each time is transportation. Cadets must review their budget and determine whether they should use POVs (private vehicles), lease a school or commercial bus and driver, or rent vans. I have observed cadet use of all forms of transportation with mixed results. Clearly the best method is a school bus and driver, but sometimes cadets have spent their budget for the FTX on other weekend activities.

Emphasize Your Strengths and Their Weaknesses

Cadets learn to lead with their strengths and match them to the weaknesses of their enemy. In the LRC FTX, the situation is a little different; however, once the mission is underway, cadets quickly learn that everyone isn't blessed with the same intelligence or physical strength. In an exercise requiring great strength to scale a 12-foot wall, it might be a small, female cadet who figures out the way for the heavier and stronger male cadets to get everyone over the wall. This station requires both skills, but not necessarily from every participant.

Set Your Timing

Usually when I review the operations plan, I find the timing of activities is not in synch with reality. For example, if you tell cadre they have a one hour and 15 minute ride to the training site, they may have the cadets leaving at 0745 and arriving when the first activity is ready to begin. They have ignored the fact that the dining hall does not serve breakfast until 0800 on Saturdays, thus they need to stop at "Mickey Ds" or somewhere for coffee and a snack; they will need a potty break once they are on site; and they should allot time for the safety briefing, which usually takes up to 15 minutes. Now they must start counting backward to determine the optimum time for leaving VMI.

Plan for Contingencies

Planning for contingencies must be a priority and not a luxury. A review of the initial mission plan in a pre-brief is a good way to begin the process of avoiding potential obstacles. In completing their plans, cadets must think about a backup response to conditions they might not be able to control. What if there is lightning or a flashflood on the course? How long do we sit it out before we go to the contingency? What if our bus doesn't show to pick us up at VMI to transport us? What if the lunch food isn't ready for morning pickup at the time we need to leave VMI? What if a training element breaks and is inoperable? One purpose of the initial briefing is to get everyone involved before the mission begins. This is the time to ask questions and to play "what if?"

Debriefs

It is an old tradition in aviation to debrief at the end of each mission. The purpose of the debrief is to deconstruct, analyze, or talk about the mission, and ask what mistakes were made. This is also an excellent leader activity for all organizations, but caution must be paid to not turning the debrief into a finger-pointing exercise. In the Air Force, the number one rule during a debrief is not to name names and to leave rank at the door. Each mission participant should feel free to speak his or her mind and to self-criticize performance. When I participated in our first debrief for Goose Creek, I knew two officers had not done their jobs properly. I was very careful not to name names when addressing the detachment commander, but everyone in the room agreed mistakes had been made and that proper correction would be taken next time to ensure mission success as planned. No one hid the fact that we could improve its performance.

Conclusion

Military life is not for everyone, and neither is VMI. Certainly, there are other much easier ways to develop leadership skills, but few colleges truly offer students an opportunity to challenge the whole self and to learn to care for others who may be struggling along the path to graduation. Students desiring a military education must want discipline and learn through the process how to self-discipline themselves. Through a Corps of Cadets Regimental System, cadets are trained how to be good followers before evolving into "sixth stage" leaders. Cadets leading a rigorous military life learn what duty, responsibility, and accountability are all about. By commissioning at graduation, many will lead our nation in times of war and serve their state in times of natural disaster and other crisis situations. Young men and women of character like them are the future of our country. More young people need to understand that our nation needs not only their skills, but also their high standards and values.

BUILDING THE TEAM
BROTHER RAT BY BROTHER RAT

*"You don't get the breaks unless you play
with the team instead of against it."*

Lou Gehrig,
New York Yankees All-Star baseball player

In 1936, the classic play "Brother Rat" became as well
known as "Cats," "Oklahoma," or "Grease." It ran a record 577
performances. Like these famous and more recent Broadway
plays, it was also made into a movie in 1938 starring none other
than our fortieth President Ronald Reagan. The original play
was written as a senior thesis by John Monks Jr. (1932) and Fred
Finklehoffe (1932). Their musical comedy was about two wild-
spirited cadets struggling to graduate after one secretly married
an officer's daughter who became pregnant. The play described a
story of the tight loyalty and close relationships fostered at VMI
and bonded by a brotherhood nurtured in both time and adverse
circumstances.

The two real life "BRs" worked in Hollywood on several
pictures including "Strike Up the Band" with Judy Garland and
Mickey Rooney. The real brotherhood of Monkss and Finklehoffe
lasted until World War II, when Monks served as a Marine Corps
major in the South Pacific. Monks continued a long career of
acting, writing screen plays, directing, and producing well-known
hits such as "The House on 92nd Street," "No Man Is an Island,"
and "The West Point Story," among others. Undoubtedly, Monks
and Finklehoffe learned the value of teamwork and the loyalty
and devotion of being true "Brother Rats."

Transformation

Producing a cookie cutter approach to leader and team development does not work. Every leader is different and each develops his or her own unique leadership style. It must be stressed that the VMI path of leader and team development is but one way to produce effective leaders. VMI utilizes what former CEO and University of Chicago professor Howard Hass calls "the tapestry of leadership." Hass uses weaving as a metaphor to explain how leadership consists of warp (vertical) threads and weft (horizontal) threads. Warp threads represent the external things in life that shape an individual, while weft threads are the intersecting strands that represent stages in life. Warp threads include gender, race, intelligence, energy, social skills, and other factors. VMI attempts to help cadets create their own distinct fiber, "starting with the warp threads of personality that are acquired throughout life."[1] In the informative stage of life, VMI takes committed students, who have a strong need to prove themselves and are willing to accomplish that goal through preparation, hard work, and the ability to synthesize data into meaningful facts and actionable plans. By accepting those committed to proving they can be successful leaders, VMI can take self-described "ordinary" people like George Marshall and give them the opportunities to develop into extraordinary leaders.[2]

At an early stage of their college education, the VMI experience forces cadets to first look inward and to discover themselves. Hass believes that leaders grow by mastering painful conflicts. Warren Bennis and Robert Thomas found that every leader in their 2002 study had undergone "at least one intense, transformational experience."[3] While these experiences do not have to be created as such and can vary in length, I believe after interviewing cadets and reading alumni comments that the Rat year is one such experience. Bennis and Thomas believe that a transformational experience is at the heart of becoming a leader. They call the experiences "crucibles." The term comes from the medieval vessel in which alchemists attempted to melt base medals in order to create gold. Today the term refers to a "severe test of patience or belief."[4] Interestingly, at VMI the transformational experience begins with Rat Crucible Day, when Rats are first tested under stressful situations. From these crucible experiences, leaders extract meaning for their lives.

While not as severe as the death of a parent or sibling, rejection by someone you love, or a divorce, the Rat year does cause a sense of alienation, which forces a "redefined sense of identity."[5] Sometimes we go through several crucible experiences. For me, there were two defining experiences of identity. The first was having polio as a child. For several years, I was isolated from others my age due to my recovery and the fear that the disease was still contagious. This experience forced me at a young age to become a much more independent person.

My second defining moment was my twelve-month tour in Vietnam and the death of a close friend from Virginia Tech who was killed in action while on patrol in the Black Mountain region near the Cambodian border. 1st Lieutenant John Hill and I were stationed in 1969 at Schofield Barracks, Hawaii prior to our deployments. I first met John six years earlier when we were both Rats in G Company, Virginia Tech Corps of Cadets. John became a platoon leader in the 25th Infantry Division. Found in his possessions was a letter to be sent home to his mother and fiancé in case of his death. In the letter, John explained why it was important for him to serve in Vietnam. John now rests on a peaceful hillside not far away, which overlooks I-81 in Buchanan, Virginia. His Silver Star speaks volumes about John and his character. While I was never seriously tested in Vietnam, I left the experience a much stronger and more confident individual. I also knew at some point I needed to be more of a servant leader and pay back the debt I owed to God for my survival.

The Rat year initially creates a feeling of separateness as cadets quickly realize that they are alone and no longer have the privileges of being a normal person. There are no cell phones to call home or cars to drive. There is limited Internet access, very short haircuts, uniforms instead of fashionable clothes, and lots of people yelling at them and stressing that they are less than human. All of the above are a part of the whirling feeling of separation.

Rats must quickly decide that this experience is what they really want and hence become followers, or they need to reach deep down inside and determine through self-discovery that the VMI path to leadership is not the path they want to follow.

For those, whose self-discovery provides a green light, these cadets soon find they share the same crisis situation with other shaved heads. To survive and thrive they must learn to be followers and members of the team or Ratline, as it is know at VMI… one Rat after another following another down a path less chosen. Each Rat "strains" (severe form of attention) and walks a prescribed line within the barracks. As followers they must learn to see the big picture and handle the small details of cadet life such as keeping their rooms clean and shoes shined. Within this controlled and closely observed crisis, Rats must have the social capacity to work well with others. If not at first, at least eventually, or they will not survive. In addition, they must have the strength of character to do the right thing and have a moral and physiological balance to pursue both personal and VMI goals. Ultimately, when Rats come together as a class, they will participate as a team in an effort that is recognized by inclusion in the VMI Corps of Cadets.

A Rat is introduced to the art of straining.
Source: VMI Communications & Marketing

The Rat year and Ratline become a transformational experience. Some hate it and become filled with anger or appear to retreat into a zombielike trance. Others become fearful. The immediate winners are those who see the experience as a big game, which like any game, the more you play, the better you get at avoiding trouble and at realizing your place and value as a pawn. The game players are the dreamers with a vision of the end goal, for they have the ability to see themselves as the future kings and queens. They are looking at their upper- class cadet role models and thinking, "Next year I can be the cadre corporal or in three years the Regimental Commander." All these emotions are natural feelings, but each Rat will handle them in his or her own way. Different tolerance levels will help determine how well the experience is handled, but with each new crisis situation growth occurs along with the personal adversity. Cadets eventually learn to cope with the trauma of daily cadet life as a Rat. Ultimately, Rats learn to convert their emotions into creativity and to look at situations from a different perspective when solving problems.

Instead of choosing the Ratline to build a network of close friends, many college students use social networking web pages such as Facebook and MySpace. Facebook has over 34 million registered users at college sites and an additional 35 million users off campus. Eighty-five percent of the users say they connect at least once a week with friends and strangers alike. It is not unusual for people to have over 1,000 "friends" in their network. MySpace is a similar social networking web page, which allows individuals to join groups and to blog on their favorite topics.

The question arising from the creation of social networks created in cyberspace is, "Do you really know with whom you are sharing personal information and can you trust these people?" At VMI cadets become a part of a social network which lasts a life time ... not just a few weeks or months. They become a part of a Rat class of approximately 425 cadets each year. The Alumni Association is commonly referred to by Virginians as the Commonwealth's biggest and best-networked fraternity. The network, composed of friendships made at VMI, lasts throughout life and each friendship has deep meaning to all Brother Rats. The network is based on people you can trust; people with integrity and high standards. There can be no better social network than to be connected to honorable people whom you can trust.

Sometimes a social network of cadets extends beyond the walls of VMI. It can include a military network of alumni or a network of former cadets from Virginia Tech, the Citadel, and VMI. One such example happened to me at Smith Mountain Lake, Virginia in 1992. I bought a Hunter sailboat from a local dealer, Dave Compton, who was a VMI alumnus of the class of 1977. I kept my boat at a local marina. One Saturday in December, I received a phone call from Dave, who told me he was heading back from Florida hauling a couple of boats. He asked me if I had pulled my boat out of the water. The temperature that night was supposed to go down to 2 degrees, and it was 7 degrees at the time and very windy. When I told him the boat was still in the water, he told me to meet him at the marina in about two hours to haul the boat out. We needed to pull it out because the boat used water for ballast. If the water froze, the hull would crack.

Later that afternoon Dave arrived with trailer in tow to help me. He still hadn't been home from his long drive from Florida. After storing the boat and draining it, he started off. I told Dave to send me the bill. He stuck his hand out to shake, and said that anyone who was a member of the Tech Corps of Cadets was as good as a VMI Brother Rat to him. Back in the day, I am not sure Dave's philosophy would have been as generously accepted by old VMI or Tech grads, who used to be arch rivals. Dave wouldn't let me pay him, and I never received an invoice. Needless to say, I sent business his way as he had personally demonstrated the extended spirit of VMI brotherhood.

Learning to Be a Follower:
The VMI Class System

Every leader is also a follower. Being a follower is a job into itself. Every person answers to someone else, an individual of higher authority. The follower's job is to do what they are told. For cadets arriving at VMI for the first time, being a follower is often difficult and many don't like being subordinate to others. Being a follower requires cadets to surrender their individualism once they matriculate into the Corps. After a time, cadets earn more independence and gain self-command or self-discipline. Hopefully, the end result will be leaders, who have learned to postpone gratification for the achievement of long-term goals.

In some ways, fraternity and sorority pledges go through a similar process of giving up their individualism for a short period to learn that they are part of a greater group with its own goals and needs. The fraternity hopes to strengthen its organization by bonding the strengths of each individual to make a stronger organization. The idea is that everyone brings something different to the table, and together they can accomplish more with their talents than as individuals. Although banned in 1885, three national fraternities, Sigma Nu, Kappa Sigma Kappa, and Alpha Tau Omega, were founded at VMI. Despite the ban, fraternities lingered on until 1911. VMI believed that while the fraternity experience attempted to accomplish the same initial goal, the fundamental difference between military colleges and the Greek system is the latter does not teach or encourage self-discipline. Self-discipline is a must characteristic for leaders.

The Greek pledging period ends with an initiation into the fraternity or sorority. Initiations are an important part of being accepted into a brotherhood or sisterhood. In the military, there is also a long tradition of initiations from special ceremonies for sailors crossing the International Dateline (180 degree longitude) to pilots flying solo and parachutists surviving their first jump. The VMI Rat experience begins with Matriculation Day in late August and ends with the ultimate initiation called Breakout. Matriculation Day is usually the third Saturday in August. New cadets arrive with their families and the bare essentials needed for daily living such as shaving gear, personal hygiene items, running shoes, laptop, and a footlocker. Students report to Cameron Hall, where they sign the Matriculation Book and officially become new cadets. They then dress in "gym dyke" (red shorts with gray T-shirt) and are given a large card with their vital information, which is then clipped to their shirts. After a family, faculty and staff dinner in Crozet Dining Hall, the pleasantries end and families depart, leaving their Rat to endure the most challenging year of their young lives.

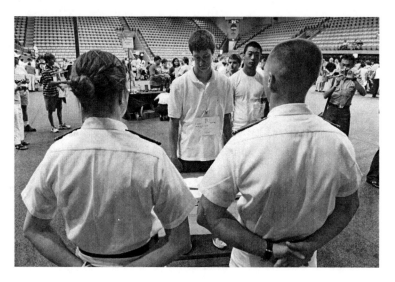

Cadets waiting their turn to sign the Matriculation Book.

Once their families are gone, cadets are marched into the New Barracks courtyard to "Meet Your Cadre." The cadet cadre is selected from the best who have "maintained strong academic records, good physical standards, and excellent peer evaluations."[6] The cadre is responsible for training Rats in their company. Like the Rats, they reported early to train for their roles. For the next few months, they would eat, sleep, run, and exercise with the Rats. The overall control and discipline of the Rats is administered by the Rat Disciplinary Committee (RDC). The overall governance of the "class system" is the responsibility of the General Committee (GC), composed of the class officers. The GC publishes the standards to be adhered to by all cadets in the barracks, military formations, and public life.

This ceremony formation traditionally brings the cadet cadre out in their gray blouses (tunics) and white pants to meet the tired, sweating Rats in their gym dyke. The ceremony is quite serious and complete with the theatrical sounds of drumbeats and flaming screams mixed with periods of silence. Spectators (faculty, staff, and other cadets) stand by the railing on the upper stoops. There is a feel of the Roman Coliseum. During this fun process, the cadre is in their face and yelling instructions in every direction. The goal is to teach Rats how to listen and to focus on exactly what is being said under pressure. To say the least, this is a stressful moment. Through this process, Rats also learn to be self-motivated. Few people will tell them that they are doing a good job. Their reward for excellent performance is the knowledge that they have done their best.

The cadre quickly introduces the Rats to the art of straining. Shouts of "tuck in your chins, suck in your guts, and thrust your chests out" are heard throughout the Rat Line on "Strain Night." Of course straining is taught to the extreme; chins and necks all meld together into one wrinkled mass of flesh. At this point, Rats are scared, tried, and just plain ugly! Once this part of the game gets old, Rats begin to learn the basics of drill and ceremony. It is amazing how many Rats under stress don't know their left from right. A simple "right face" command may end up with two Rats touching nose to nose.

Rat focuses on reading his *Rat Bible.*
Source: VMI Communications & Marketing

Sunday is Rat Crucible Day. For a solid week before classes, the new cadets learn the VMI way of conducting their daily lives for the next four and sometimes five years.

Rats are also taken to the Parade Field and taught how to march and to do the manual of arms with their weapons. Anytime there is spare moment or rest period while training, Rats are expected to study their *Rat Bible*. This usually creates an unusual scene, because Rats hold the *Rat Bible* about six inches from the tip of their noses and sit or stand at attention while memorizing various facts. Sections of the *Rat Bible* include special orders, names of cadet regimental staff, names of New Market dead, and much, much more. Memorization is also a technique used to teach new cadets traditions and to test them under pressure. With time, the more they recite, the more focused they become.

Rats are also taught during their first week how to keep their rooms in tip top shape and to have them inspection ready on a moment's notice. Because Rats share a room with up to four others, they quickly learn the meaning of teamwork. If one Rat doesn't do his or her share, whether dusting and mopping the floors to cleaning the sinks, everyone in the room gets demerits and pays the penalty. They quickly learn the true meaning of responsibility and accountability.

Another important phase of Rat training is physical fitness. Rats spend a lot of time running. They run in gym dyke and they run in ACUs with combat boots. In the first few weeks, it is not unusual to see a few Rats on crutches. Most are hurting from sprains or blistered feet and total exhaustion. The new desert tan suede combat boots issued to cadets are far easier on feet than the traditional stiff leather army boots, but lots of running still causes lots of blisters. To prepare Rats for the first few weeks of running, cadets are encouraged to run in their boots at home and to use mole skin and other foot products to protect their feet. While physical fitness exercises and running aren't tons of fun, evenings are full of competitive athletics such as flag football, soccer, and other activities.

After a week of intense training, the Rats are ready to meet the Corps. Late on a Sunday night before classes start on

Tuesday, the Rats are herded out of their rooms to begin "Hell Night." Traditionally, Rats are escorted at a jog across the street to Cocke Hall gymnasium, where they meet the returning First Class cadets. These veteran cadets then put the Rats through their paces. Rats are everywhere doing pushups, running, flutter kicks, and a variety of other physical activities. Sweat parties are now limited to 15 minutes. There are four sets of three minute exercises, each followed by a minute of rest. At the end, the Rats are marched back across the street to the New Barracks, where they are introduced to the General Committee and given a traditional speech by the Regimental Commander.

Rat, Nash Roberts (2011), undergoing
physical training workout.
Source: VMI Communications & Marketing

After two weeks at VMI, Rats are bused to Harrisonburg where they begin a march to New Market, location of the Civil War battle where 10 VMI cadets were killed and 47 wounded in action. VMI cadets are the only cadet corps to serve as a unit in combat. Two-thirds of the deaths on May 15, 1864 were Rats. Once at the battlefield, new cadets are given an orientation by upper-class cadets on the battlefield. The importance of New Market and the sacrifices made by members of the Corps are explained. New cadets are led in a reenactment charge by upper-class cadets across "the Field of Lost Shoes." At the end of the orientation, new cadets line the Bushong Farm Lane and greet First Class cadets, who have marched the 80 miles from VMI to New Market to set the example for the new class of Rats. The First Class cadets carry shoulder boards for the Rats, which are pinned on in a ceremony where the new cadets are sworn into the Corps. The Ratline is now complete.

During the event the Rats will take their oath of cadetship:

> *I hereby engage to serve as a cadet in the Virginia Military Institute for the term for which I have entered. I promise on my honor while I continue a member thereof to obey all legal orders of the Constituted Authorities of the Institute and to discharge all of my duties as a Cadet with regularity and fidelity. I will never lie, cheat, steal, nor tolerate those who do, and I solemnly pledge to keep this covenant with all members of the Corps, so help me God.*

The following weekend, the Rats are bused over the nearby Blue Ridge Mountains to the town of Bedford and home to the $24 million National D-Day Memorial. It is here at the memorial that cadets once again are reminded of the importance of sacrifice, commitment, and duty. During the invasion of Normandy, France, 21 men from Company A, 116th, 29th Infantry Division were killed, 19 on the first day. The town of Bedford had the highest per capita losses of any community in the U.S. This beautiful and unique memorial stands in tribute to the 150,000, who risked their lives and followed their leaders onto the beaches of southern France to free Europe from the grips of Nazi Germany.

Rites of Passage

During the academic year, Rats are encouraged to be inventive and to come together as a "Rat Mass." This often leads to class pranks, which inevitably lead to some form of punishment by the commandant's staff or the Rat Disciplinary Committee (RDC). More likely than not, the punishment is administered by RDC. Wayward cadets report each evening to the RDC at 2230. While the location varies, Rats are presently led to a room on the fourth floor stoop, where they are accused of violations ranging from not showering to failure to shine shoes … all trivial offenses. The RDC has no authority over any serious violations. Their job is to enforce the written and unwritten codes that govern day to day life. While waiting to be heard, Rats are required to exercise. After hearing the Rat's case, the Rat is required by the RDC to return to hear their punishment the next evening.

Various forms of hazing became popular as early as the 1850s and were used as a form of punishment. In 1858, records indicate the first dismissal of a cadet for abusing another. At VMI hazing once required cadets to lie across a table where they were hit by broomsticks and sword scabbards. Despite State Attorney General William H. Richardson's protests, it continued. Richardson felt hazing brought more "discredit and injury" to VMI than anything else. He called hazing "unmanly and cowardly."[7] One Jewish cadet, who was reportedly hazed the most, was world renowned sculptor Sir Moses Ezekiel (1866) of Rome, Italy.

Hazing was expected to be taken by cadets. Serious injuries were reported as accidents of friendly scuffles. In the late 1890s, George Marshall established a name for himself by refusing to report cadets, who hazed him to the point of injury, when he fell onto a bayonet after being required to squat over it in a strained position. Superintendent Scott Shipp was so appalled by the unbridled hazing that many believed it finally led to his retirement. In 1918, the reported practice was called "laying on of the hands." It was abolished by Superintendent Lieutenant General John Lejeune, former Commandant of the Marines. During his time, cadets bragged, "give us your son, and we'll give you back a man."[8]

When I was "turned" (Virginia Tech version of Breakout) after nine months of being a Rat at Virginia Tech, I and my fellow Rats in G Company were initiated (hit usually) by each upper-class cadet in our company. To my best recollection, I was hit approximately 78 times with booms (weapon of popular choice), razor cords, leather belts, fraternity paddles, and coat hangers while bending over and touching my toes. Some, fortunately, were light taps and were meant to scare more than hurt. All of this happened in a one-week period. Regardless, I was black and blue from my waist down and picked broom bristles from my rear end for a week. If that wasn't bad enough, when I was inducted into the VT Monogram Club, future lettermen were blind-folded and paraded through the outdoor quad in our under shorts and then told to bend over. You get it, more of the same, except my paddlers this time included some very large football players . . . but I earned their respect. Wasn't that the idea? Somehow, I don't remember anything in these experiences that made me a man or a leader, but they make for some great "old corps" stories.

After a landmark year of physical hazing (over 40 cases reported), a storm of complaints, and an inquiry by the Governor Harry Byrd in 1927, a commission recommended cutting off funds to VMI. The governor demanded that hazing be stopped, but in general he supported the Institute. Legislation was then passed in 1928 with broad support of the recommendation; however, it did not define hazing as other than physical abuse or injury (vaguely defined) and left the door open to emotional and mental hazing. Even then it continued. VMI history reports that "severe" hazing (physical abuse) stopped before the 1980s, but until various states started a crack down on fraternity hazing during the 1980s and 1990s (39 states had legislation by 1998), it still existed in military colleges and Greek systems everywhere as the method of passing one into brotherhood.[9] In March of 2003, Virginia revised the old law and tightened the definition of hazing.

Unfortunately, according to a 2008 study by the National Association of Student Administrators, hazing still exists today, but in a different form. Over half of the social, academic, and cultural clubs on U.S. campuses reported hazing new members. The most reported type of hazing is forcing new members to drink excessive amounts of alcohol. Seventeen percent of men and nine percent of women reported drinking to the point of passing out or getting sick.[10]

Fortunately, at VMI abusive physical behavior will send you packing your bags faster than just about anything. Physical hazing is not an honor offense, and thus there is no trial. Goodbye and good riddance! While alcohol abuse is a major problem on all campuses, it is generally handled in a different manner at VMI. Administrative efforts focus on education and other techniques to stop this common practice.

Today, Rats are initiated into the Corps with a ceremony better known as "Breakout." While the timing of Breakout has varied over the years, currently it is held around the end of January; however, Rats are never told the exact date or time. Thus, rumors run rampant through the Corps all year as to when Breakout will occur. The week before Breakout is called "Resurrection Week." During this week, Rats are commonly seen on the drill field doing heavy doses of physical training. Over the past 25 years, Breakout has consisted of running a gauntlet from the ground-level center courtyard to the fourth-floor stoop while being pounded; climbing a muddy hill doused in water with upper-class cadets pushing Rats backwards; enduring heavy physical workouts; and in 2008 it consisted of returning to New Market via a 12-mile march complete with Alice packs filled with 30-35 pounds of sand plus rifles.

Rats undergo a full day of physical activity before
being initiated into the Corps during Breakout.
Source: VMI Communications and Marketing

The Saturday morning also consists of a full day of very physical and ceremonial activities. As a casual observer can ascertain, each year Breakout becomes more professional and is managed in a more controlled and meaningful manner. After New Market, cadets are then bused back to Post and are treated to a steak dinner with their dykes. Finally, in an emotional ceremony around the courtyard sentry post in the Old Barracks, the Fourth Class is welcomed into the Corps of Cadets. The Fourth Class is now officially a part of the Corps team.

Building a Successful Team

Teamwork is a very simple concept. If we work together to maximize our individual strengths and minimize our weaknesses, it becomes possible to maximize performance beyond what the individual members of the team could accomplish. No one understood the meaning of teamwork better than George Marshall. When Marshall was assistant commandant of the infantry school at Ft. Benning, he began the process of building a future leadership team. Marshall understood that to be a great leader, you needed a team of loyal followers. In 1927, he developed a process of identifying people, whose abilities as leaders or exceptional followers could assist him in case of a future war. Marshall's list contained the names of over 200 officers, who during the war were promoted to the rank of general and given command or staff positions of significance. These officers became his team, which led the victory in World War II.

VMI also operates as a team. As a team the Institute attracts young men and women, who work hard, perform well, and are often as not high potential, over achievers. They push themselves to be a part of this unique and prestigious college. Challenge stimulates and drives them to higher levels of performance. In the process, they learn to do things they never dreamed they were capable of doing. It is these challenges that create self-confidence.

When building their team, the Rat Mass does not have the luxury of selecting their ideal BRs or cadre to create the ideal class of future graduates. All cadets involved must work with whom the admissions staff enrolls and those assigned to their squad, platoon, and company. Some cadre members may take a look at their Rats and decide "they don't belong at VMI." Perhaps their Rats are too heavy, too skinny, or they think they have an attitude. For whatever the reason, this does occasionally occur; however, when it is apparent that cadre members are actively encouraging cadets to leave, this behavior is quickly squashed by the commandant's staff. In cases, which are more than the typical homesickness, Rats are encouraged to see a counselor. In addition, cadets resigning must sign a statement which asks whether their resignation is because of physical abuse or being forced out of the Ratline by upper-class cadets.

George Marshall actually received comments that he didn't belong at VMI from his own brother Stuart, a 1892 graduate. Stuart was horrified that George wanted to attend VMI. He told his mother that George would "disgrace" the family. This single comment made Marshall more determined than ever to try his best and to excel on every front possible at VMI. The comment also eventually estranged Marshall from his brother, and George even went as far to marry Lilly (first wife), a Lexington girl, whom Stuart had also courted as a cadet. Sheer determination has always been a defining trait of VMI graduates. Marshall was no exception.

In a real world setting, cutting low performers is thought of as hard nose leadership and something required to let teammates know that standards are important and will be upheld. We see examples everywhere in business and the military. Knowing only the best will become Army Rangers or Navy Seals is part of what makes these units special. Even corporate giant General Electric (GE) tracks its leaders through constant performance evaluation. GE identifies the bottom 10 percent and then makes an effort to develop those individuals into better performers. Those not rising to the standard are cut from the workforce.

In many ways, VMI also does the same thing. They just do it differently and allow the Rat more time to make the standard. The standards at VMI are relatively simple: maintain a grade average of C or better in a rigorous curriculum; take and hopefully pass a challenging PT test each semester; stay as demerit free as possible by obeying cadet regulations (*Blue Book*); and follow the strict honor code. Many do not pass one or more of these standards, and thus they do not earn the coveted VMI diploma. But others may stumble academically in the Rat year and come back charging to graduate with honors or as outstanding Corps leaders. At no time should a cadet be run out of the Corps due to race or gender.

An emphasis area for business schools is teaching students to learn how to work in teams. The ability to work in teams is essential in the business world and the military. Generally, students hate teams because of the amount of "social loafing" that occurs. Often one or two students do most of the work, while others do little or produce poor quality work and earn the same grade. The opposite is having a team of high achievers, who spend most of their time in-fighting over who has the best ideas. High achievers often make the worse teammates as can be noted on basketball teams. High paid stars sometimes refuse to play team ball, because they think they are better than their teammates. They want the credit for victories and blame others for team failures. These individuals have not learned the importance of respect for others.

Respect is essential in building teams of character. Team members must not only respect their leader, they must respect their fellow teammates. For a marketing project in 2008, my cadets participated in a Google ad campaign competition. One group (who failed to function as a team) had major problems. The members of the team did not respect the leader, and the leader failed to show them respect after he "flamed" them in a classic out-of-control email. Major Elizabeth Baker, our MIS professor and team co-sponsor, and I spent considerable time counseling these cadets on what it meant to be a team member.

In similar academic situations, I observed teams where zealous, grade fanatical students hijacked a team and refused to delegate the work load. These students failed to understand that t-e-a-m does not spell "me." They did not understand the concept of being a good follower. As an instructor, I encourage students to share the leadership role. I do this by ensuring that they attribute all work that they do on a paper or project. This way they are held accountable for their work as well as the overall team performance.

To see an application of the team concept in action, we have to look no further than one of my favorite grocery store chains, Whole Foods Market. At this chain, the basic organization unit is not the retail store, but the team. Most stores have approximately eight teams that oversee various departments from the bakery to checkout. Every new employee is placed on a team, and at the end of four weeks his fellow team members decide the candidate's fate. A two-thirds favorable vote earns a spot on the team. This selection method is also used for positions at company headquarters in Austin, Texas. These small teams are responsible for all key operating decisions at each of the 265 stores.[11] Apparently this model of leadership and management works. In 2007, Whole Foods Market was ranked number five in Fortune's "100 Best Companies to Work For."

At VMI, Rats are also taught to share the load by a rotating a posted schedule of duties and responsibilities for which they are accountable. Each of these duties is spelled out in the *Cadet Handbook*. When team members work together with equal load and leadership sharing, they will perform better. They develop a sense of team worthiness. As success is achieved, recognition for a job well done is slowly provided in simple words or phrases. In the end recognition is won by being a member of the best platoon or company in the battalion or Corps

Team building must include some fun and activities that develop esprit de corps. In an effort to prove their unity, class members often work to create unique class mottos and symbols to develop pride in their class, team or unit. For example, many class members work on the Ring Committee to capture a unique symbol of their class. While there are few cadets at any military academy or college that will ever admit to having fun, there have to be enjoyable moments in the process. Otherwise, I can't explain the many smiles I see each and every day at VMI. Furthermore, the retention rate would be much lower than it is.

Of course, what is fun can only be defined by the individual. Many cadets find fun in the company level, intramural sports program. These co-ed sports such as flag football produce many a memorable moment as cadets compete for battalion and Corps championships. Many cadets think its lots of fun to be bused to Washington D.C., Richmond or Williamsburg to see the football team play their rivals or to march in the Presidential Inauguration (13 times since Howard Taft's inauguration in 1909), New Orleans's Mardi Gras, or the Pasadena Rose Bowl (2008). To others, Corps trips are a big hassle of long rides, lots of waiting around, and too many restrictions.

Pranks are also perceived by some as fun and a sign that Rats are ready to breakout. In another era, one group of cadets managed to sneak out at night and placed the barrel of a cannon through the first floor window of an upper-class cadet room. Most recently, a Rat class locked themselves in their rooms and barricaded the doors so they could avoid a sweat party. In my case at Virginia Tech, the 32 rats in my company felt it would be "fun"

President Barack Obama reviews the
VMI Cadet Corps at the 2009 Inauguration Parade.
Source: Department of Defense

in the middle of the night to abduct an obnoxious Third Class
cadet, throw him in the showers, shave his crotch, and paint our
class numerals (68) with shoe enamel on two personal parts of his
body. Everyone in the Corps seemed to think that act was terribly
funny, but the commandant thought it was just terrible. Later as
I served my six weeks on social probation, I quickly forgot what
had been so much fun.

Regardless, a team builds cohesion by participating in
activities, where laughing and crying together is a part of the
experience. Somewhere in the process, you also learn it is okay
to tease your Brother Rats, create nicknames for them, support
them, and sometimes tell them they are crude, rude, and sociably
unacceptable. Isn't that what brothers and sisters do?

Rats become family, and members of a family must care
for each other. Thus, the final element of teamwork is caring
for your teammate. A teammate might literally be your right
arm in a challenging situation. A cadet's relationship with

others on the team can make a huge difference in the team's success. Relationships begin with listening. How can teammates understand the importance of picking up the slack for their BR, if they are not listening? An individual might be depressed, in need of academic help, sick, just lost a girl or boy friend, or worse yet experienced the death of a loved one. I am always amazed at how some people can go through the year knowing a person is experiencing a difficult time and never acknowledge the fact by saying something as simple as "Is there anything I can do to help?" or "I am so sorry to hear about your unfortunate news." Of course a personal card or email is also appreciated.

By ignoring someone's pain, a message is sent that says "I really don't care about you." As team members, we need each other, if not today, tomorrow. Brother Rats learn by example the importance of being there when it counts, whether marriages or funerals. Caring for others also helps create a cohesiveness, which in turn becomes a force multiplier allowing the team to do more than would be expected in times of trial. Caring must go beyond cadets, for everyone at VMI is a part of the team.

Mentorship: The Dyke System

Mentorship and coaching are two important elements of a leader development program. In the business world, mentoring is often an "arrangement in which experienced employees assist less experienced employees grow and advance" in their careers. Mentors assist by providing advice, support, and encouragement.[12] Cadet mentoring has a long and storied history. At VMI, it is called the "dyke system." The purpose of the dyke system is to ensure that Rats receive one-on-one assistance in as non-threatening a way as possible. Most upper-class cadets take this responsibility very seriously; however, some are known as "Rat daddies" for their soft approach to handling Rats. More than often, the dyke's room becomes a sanctuary for a Rat and the two are on a first name basis. This relationship is usually shared by other dykes in the same room. The dyke's roommates are known as "uncle dykes" to the Rats.

At VMI, an astute observer quickly notices that all Rats are not created equal. Some Rats, like some employees, need more attention and help than others. Cadet mentors, who are First Class cadets, can be especially helpful in assisting less experienced cadets learn the ropes of the VMI culture. They teach them the practical shortcuts that can make their life much easier. For example, what daily chores does a Rat perform that could be done every other day? How can those chores be done in a more efficient and effective way? What are the best ways to keep upper-class cadets off your back? How do you win "the game?"

In a good mentorship arrangement both parties find the arrangement beneficial. The mentor often gains an ear to the feelings and attitudes of the younger generation or in this case the Fourth Class. For example, are there specific problems with certain upper-class cadets who are illegally hazing, out of line, or acting in a non-professional manner? Is the Fourth Class planning an activity, which is not in their best interest? Is a Rat in academic trouble and in need of some tutoring? In turn for providing this valuable teaching and mentoring, Rats provide small services for their dykes. "Cross dyking" is not permitted, or in other words, male dykes cannot be assigned female Rats and vice versa. It is the Rat's responsibility to do nominal chores for the dyke. For instance, Rats might keep their dyke's room clean, roll their hay, polish their brass, pick up their laundry or mail, or shine their shoes. On weekends, they may watch movies or play video games in the dyke's room.

Most dykes are assigned, but often these arrangements are made prior to a Rat ever matriculating. In these cases, the upper-class cadet and the Rat have something in common such as they graduated from the same high school, are from the same hometown, will play on a varsity team together, or are possibly friends with an older sibling of the Rat. Regardless, the dyke system is not intended for the Rat to have an upper-class buddy or a tormentor if chores are not completed. The task of the First Class cadets is to educate their Rats and to develop a relationship, which becomes the Rat's support system while at VMI. Visible friendships should come after the dyke's graduation not before.

After graduation, these friendships often last a lifetime, and it is not unusual for dykes to attend the graduation of "their Rat." For many, the Rat and the dyke are like an extended family; and like family their relationship will have ups and downs. As in the real world, cadets must learn to balance the relationships in their lives whether work or family.

In the business world, mentors often select their protégés and then begin a process of unofficially guiding them through the hurdles of their career journey. One well known example from VMI lore involves George Marshall and his relationship with General John J. "Blackjack" Pershing. While in France during World War I, Pershing became very impressed by this junior officer, who would stand up to him and give straight accounts of battlefield situations. By the war's end, Marshall was serving as Pershing's aide-de-camp. This direct relationship with Marshall serving as Pershing's aide continued until 1924. Throughout Marshall's career, Pershing served as his mentor. Interestingly, in World War II it was Marshall's time to serve as a mentor. Marshall chose to mentor Dwight Eisenhower, a man who continued to ask Marshall for advice even through his presidency of the United States.

Coaching the Team

Coaches on the other hand offer more direct and immediate feedback when improvement is needed. At VMI the cadet cadre acts as coaches. The cadre is expected to tell the Rats "what right looks like." These cadets are there to model positive behavior and to make on the spot corrections and to ensure that the correction has been made. This is accomplished by providing what Lieutenant General Robert Flowers (1969), calls "eyeball to eyeball" leadership. It can only be done if you are with your troops or cadets and not leading from the desk. Flowers said in his retirement speech that he regularly ran with the troops and talked to them one on one in informal settings.[13]

One key aspect of coaching is to differentiate between mistakes of commission versus mistakes of omission. If a Rat or less experienced cadet makes an honest mistake which reduces performance, then this is a mistake of omission. Leaders should quickly correct such mistakes, but move on and focus on lessons learned. Mistakes of commission, however, are violations of rules and regulations or the values of the Corps. Like in a business situation, the cadet should be punished, and if in a leadership role they should be removed.

Cadet leaders must learn how to make appropriate corrections when in the coaching role. When I was a Rat, the Corps marched before the reviewing stand, when it was noted by an evaluator that my M-1 rifle was canted too far left, and thus our company earned a lower score for the parade completion. When we returned to the barracks, I was summoned for a royal chewing out. My punishment was to sleep with my fully disassembled M-1 for a week. Another cadet at summer field training at West Point left his guidon unsecured. It was picked up by a cadet cadre member who counseled the cadet. The cadet was punished by being forced to carry a miniature 12 inch guidon (made by the cadre member) everywhere (including bed) for a week. Were these appropriate punishments? The Army captain who was responsible for the cadet company thought otherwise and initially counseled the platoon leader for his questionable judgment. Later he gave him an A for creative coaching and supervision.

Nowhere is coaching and teambuilding more evident than during Rat Challenge activities. Rat Challenge was created in 1968 by Dr. Clark King, the beloved assistant football coach and head of the Physical Education Department in the 1960s and 1970s. The purpose of Rat Challenge is to foster self-confidence and physical conditioning in Fourth Class cadets. It was also designed to give First Class cadre leadership opportunities. Aside from Corps regimental cadre, Rat Challenge has its own cadre; thus, the event provides an additional leadership opportunity for cadets. More on Rat Challenge is discussed in the Chapter 9, "Never Say Die."

Fairness and Equality for Team Members

There is no question that the VMI culture has been shaken and transformed by diversity. For most of its existence, VMI cadets could be described as members of the "lucky seven."[14] Most cadets were male, young, white, Christian, heterosexual, able bodied, and middle class. A critical mass of research offers evidence that those, who are not members of these groups, tend to "suffer disproportionately from out-grouping and marginalization in American society."[15] Prior to the 1960's, admissions standards denied enrollment to members of several categories. Some critics believed that VMI graduates possessed an unfair advantage in the state's hierarchy of leadership roles and influence. The question became whether anyone from the "unlucky seven" category would want to gain admission to the Institute?

The question was answered in 1968 with the admission of African-Americans and in 1997 with the admission of women. The answer was yes, but the decision to admit these groups was hastened by the Federal legal system. To balance the playing field at VMI and other colleges, the U.S. Supreme Court intervened and legally upset the status quo at the Institute. The first intervention affected all colleges and universities. The U.S. Supreme Court ruled in 1954 that schools must desegregate. On August 21, 1968 five African-American cadets matriculated to VMI. This admission of five new cadets was 14 years after the Supreme Court's landmark decision. The five pioneers were the first to apply and meet all admissions requirements. The Institute did nothing to block their admission or to discourage their applications.[16]

In the beginning (1968-75) most African-Americans admitted were scholarship athletes. VMI had little financial aid at that time to offer any cadets and the best academic students were going to the academies and Ivy League colleges. With the leadership of Cadet Frank B. Easterly, 1969 class president, the addition of African-American cadets went smoothly. Easterly made it clear through class meetings that no forms of discrimination would be tolerated by any race. Every cadet, once admitted, would have to earn his "stand" in the Corps and all would receive equal treatment. Of the first five admitted, three did "exceptionally well." Of the other two, one drowned in an accident on the Maury River and the other dropped out.[17]

To assist high-risk, but capable, minority males gain admission to college has been the mission since 1987 of the privately funded COW (College Orientation Workshop). Under the founding leadership of Gene Williams (1974), the COW program has fashioned a unique VMI approach to preparing students academically and physically for the rigors of college life. The four week program offers students "immediate feedback (positive/ negative), brutal honesty tempered with compassion, continuous mentoring, demonstration of sincere concern and interest, and uncompromising high expectations."[18] Over its 20-year history, more than 70 percent of its 500 plus graduates have gone on to attend college. To date more than 15 of these students have matriculated to VMI. One VMI graduate, Richard Borden (1993), credits the program with saving his life after once headed down a different path in the Bronx. "VMI and COW helped give me a fresh start, where I could focus my talents and leadership on positive things."[19] Borden is now a successful naval officer.

One of VMI's many successful minority military officers is Major General Darren McDew (1982). McDew, a former vice-commander of the 18th Air Force at Scott Air Force Base and military aide to President Bill Clinton, is presently the Director of Public Affairs for the Air Force. He is the eighth VMI graduate to obtain flag officer rank in the U.S. Air Force, and VMI's highest ranking African-American in the U.S military. His career began as a second lieutenant piloting a KC-137 refueling tanker. McDew earned the rank of general after 23 years and appears to be on a fast track to further promotions and commands. His past responsibilities with the 18th Air Force included supervision of a 54,000 military and civilian work force. The 18th Air Force is the war fighting component of the Air Mobility Command. McDew also recently served a tour as CENTCOM director of mobility forces.[20]

In the civilian sector, Anthony Q. McIntosh (1989) made his mark with Motorola. McIntosh is currently director of Global

Product Marketing for Motorola's Mobile Devices Business unit. At VMI he was a Distinguished Graduate with a double major in economics and modern languages. He also graduated as a Distinguished Military Graduate and served as an officer in the Corps of Cadets. McIntosh was a four year football letterman, and he earned two letters in track (All-Conference). After VMI, he worked for CSX and Frank Lynn & Associates consulting and later earned his masters in management from Northwestern's prestigious Kellogg Graduate School of Management.[21] He currently is serving a term on the VMI Board of Visitors.

Also serving on the Board of Visitors is Michael W. "Max" Maxwell '87. This successful graduate is vice-president, Safety and Strategic Services for Pepco Holdings, the parent company for Potomac Electric Power. While at VMI, Maxwell was a four-year letterman, team captain, and Southern Conference Champion in indoor and outdoor track. After graduation, he served eight years in the U.S. Army Reserve; served as a member of the Board of Directors of Boys and Girls Club of Greater Washington; and was a member of the Leadership Greater Washington.

Another successful minority, Darryl K. Horne (1982), is also a member of the Board of Visitors. Horne was the founder of Horne Engineering, a firm which merged in 2005 with Spectrum Sciences and Software Holdings (350 employees). He is now the president and CEO of Spectrum Holdings. After earning a degree in civil engineering, Horne served in the Army Reserve as a captain. In 1999, he was named by Ernst & Young as a Greater Washington Entrepreneur of the Year and again in 2002. He was also a finalist for a National Capital Business Ethics Award. In addition, he serves as a Trustee of the Federal City Council, a "non-profit, non-partisan organization dedicated to improvement of the nation's capital and composed and financed by the region's top business, professional, educational, and civic leaders."[22]

During my tenure at VMI, I had the pleasure to work with many minority cadets and staff. At my first reception at the superintendent's house, I met a Corps legend, retired Sergeant Major Al Hockaday. The Sergeant Major first came to VMI as a NCO instructor for the Navy-Marine Corps ROTC program.

He was one of the first and few African-Americans assigned to VMI by the Department of Defense to instruct cadets. The cadets loved this highly decorated Vietnam veteran, and he became the Cadet Corps sergeant major after his retirement from the Marines. Hockaday and his wife now own and operate two downtown specialty stores frequented by tourists and locals alike; however, he still shares his time and energy with VMI cadets. Presently, he is helping with the Character Counts program by assisting cadets present lessons on core ethical values (trustworthiness, responsibility, respect, fairness, caring, and citizenship) to local school children. His professionalism and pleasant personality have earned him the respect of all. Sergeant Major Hockaday and other NCOs who serve at VMI shape the lives of all cadets.

Five of my outstanding minority cadets were Jamal Stafford (2004), Tamara Ferguson (2004), Charles Newke (2007), Sam Avarenga (2007), and Joe Taylor (2009). Stafford worked with me in 2003 and 2004 through the LeadAmerica organization as a team and site leader at the Goose Creek Outdoor Education Center. As a student he took home many academic honors (wore academic stars every semester) and graduated from the University of Virginia Law School in 2008. His quiet style of leadership offered encouragement to many students. Stafford was an outstanding role model.

Charles Newke, Joe Taylor and Sam Avarenga (2007) were serious students and athletes. Newke was named Big South Conference Freshman of the Year for his performance on the soccer field and was consistently an All-Conference player. In 2007, Newke received several tryouts for professional teams including the Columbus Crew. As a student he was always a very thoughtful thinker and had a sincere desire to improve. Newke played minor league soccer (USL) for the Wilmington Hammerheads and now works as a manager for Norfolk & Southern. His brother Tony Newke (2010) is also an economics and business major and on the soccer team. Almost a dead ringer for his older brother, Tony is also a thoughtful, student-athlete.

Avarenga was an outstanding wrestler. He served as the team captain in 2006-2007 and earned Big South Conference honors for his excellent record of wins and tournament victories. Joe Taylor was a hard worker and his efforts showed in the classroom and on the football playing field. He was well respected by his fellow cadets. On the field as a linebacker, he earned "Big South Defensive Player of the Week" honors, and he was frequently on the Dean's List. Tamara Ferguson, a native of Canada and a sprinter, was a valued leader and team captain of the women's track and field team. As president of the Promaji Club (African-American service organization), she led efforts to assist with Project Horizon (local women's shelter) and to landscape Lylburn Downing Middle School. After graduating, Ferguson was employed by Cintas (world's largest uniform company) as a manager trainee.

The admission, acceptance and integration of women into the Corps was not an easy transition. First many VMI alumni and members of the Board of Visitors strongly opposed the admission of women. The battle started in 1989 and lasted seven years. To fight the case filed by the Justice Department, the VMI Foundation spent $10 million in legal and consulting fees from its endowment. The case was filed by an anonymous female who was denied admission because of her gender. From 1988-1990, 347 other women also applied for admission. VMI won the first legal round in U.S. District Court, but lost the following appeals.

To head off the legal challenges, VMI, in partnership with private Mary Baldwin College, founded the Virginia Women's Institute for Leadership (VWIL). Thus, a female cadet corps within the civilian single-sex (female) population of Mary Baldwin College was created. The VMI Foundation funded the program, which is still in existence and thriving under the leadership of Brigadier General Mike Bissell, one of VMI's most highly decorated veterans and the former VMI commandant of cadets. Finally, on June 26, 1996 the U.S. Supreme Court ruled that VMI could no longer remain a single-sex college. The alternative was to go private and accept no public funding for its programs. In effect, VMI would have lost its highly prized ROTC programs and thus its ability to commission officers.

General Bissell was up to the challenge of developing an all women's unit at Mary Baldwin (VWIL). As a former Harvard Fellow, director of flight training at Fort Rucker, and program manager of the Boeing/Sikorsky stealth helicopter, Bissell developed, organized, and implemented the nation's only female corps of cadets. In 2006, Bissell was inducted at Fort Rucker into the Army Aviation Hall of Fame. The Medal of Honor nominee was awarded the Distinguished Service Cross for his heroism while attempting to rescue a "severely wounded" advisor from a hot landing zone (LZ). On his second attempt his aircraft was hit by hostile machine gun fire, killing his door gunner and wounding Bissell in the neck. On his third attempt he was successful, but the damaged aircraft was forced to crash land after losing its engine.[23]

The Supreme Court did not accept the concept that VWIL offered an equivalent leadership program for women. Needless to say, the court decision made national headlines and created many emotional responses. After a September 21, 1996, vote by the Board of Visitors, it was official: VMI would accept women beginning the fall of 1997. The delay gave VMI an additional year to prepare. The administration did not want to act in haste, as many felt The Citadel did when it admitted Shannon Faulkner, who left after several days and caused a public relations nightmare for the Citadel.

Once ordered to admit women, Superintendent Josiah Bunting III and his staff assumed women would be admitted, and they began a comprehensive planning process to address every foreseeable contingency. Bunting declared that the "assimilation" process would be completed in an exceptional manner. He and others did not want a replica of the Citadel fiasco. To ensure that things were done correctly and not in haste, the Institute lobbied for state funding and was awarded $5.2 million for the transition and necessary construction projects.[24]

A campus committee was formed to determine what VMI valued as an institution, and what was necessary to maintain the integrity of the VMI experience. Without over simplifying the long and dedicated hours spent on this endeavor, Bunting's executive committee decided to keep such time valued elements as the honor system, the Ratline, short haircuts, the dyke system, physical fitness standards, curriculum, and uniforms among other important components of the VMI experience. Thus, it was decided early on that everyone attending VMI would have the same experience. There would be no letting down of standards, or female cadets would not earn the respect of male cadets. The adopted philosophy became "one standard, one Corps."

Of course there is much more to the story of female assimilation into the Corps of Cadets. Laura Brodie (Ph.D), the band director's wife, did an outstanding job of detailing this painstaking task and the problems VMI encountered. Her 2000 book, *Breaking Out*, was well received on campus and received critical acclaim for her balanced and fair assessment of the process and the Institute. By the time, I arrived in 2002, the controversy had settled down and most assimilation mistakes had been recognized and corrected.

Female cadets have been fully assimilated into all aspects of the Corps. This group of Cadets are being commissioned as 2nd Lt. in the Air Force.
Source: VMI Communications & Marketing

My first year, I discovered just how determined VMI women were to be treated equally. Much to my complete surprise, I was confronted after class by a young woman who was a First Class cadet. She said to me one day, "Sir, I am calling you out." I had no idea what she meant. She informed me that I was calling on her more than certain male students, and that a few male students were keeping a count and laughing at her every time I called on her for responses to my questions. I guess there are several ways of looking at this, but I knew I had never been accused of treating female students any differently than male students. I had heard throughout my career that studies indicate that in many classes, male professors do not call on females and in other cases females were less likely to respond to open questions from male professors.

Now that I was teaching at VMI, where potential claims of sexual discrimination and harassment ride the surface of cadet life, I wasn't about to permit a female cadet be intimidated by her male classmates (or so I thought) into not participating in a discussion-based course. Since I didn't keep a running account of how many times I called on cadets, I had to assume she was correct. Thus, I backed off and encouraged her to answer questions on her own if she didn't want me calling on her. To cut her a break because she was a female would have been treating her unequally. All I expected from her or any cadet was participation in class discussions. Regardless of the outcome, I did become more conscious of how I handled participation issues with all cadets.

In numerous observations, I noticed that many female cadets go out of their way to be one of the "guys." Most female cadets do not seek or want any extra attention because of their gender. They truly want to be treated equally. One female cadet responding to a *Richmond Dispatch* article on the 10th anniversary of women in the Corps responded, "the females that are cadets at VMI … most try only to keep their heads down and make the best of it like the guys do. This article draws unwanted attention to them and makes it only harder for them to keep a low profile. Is there going to be an article for every anniversary of the decision?"[25] The answer is probably yes as long as there are still alumni and male cadets who feel women don't belong in the Corps. Unfortunately, there are still some who fail to see the gains that VMI has made by admitting women.

Regardless of feelings some may have about women in the Corps, I am sure they all love my favorite illustration about women at the Institute. It was also a favorite on YouTube.com and in the press. During the 2006 VMI-Citadel football game, several Citadel rats came over to the VMI side to steal the VMI flag. Immediately, the VMI Rats ran to defend and retrieve their broken flag. With the event captured on TV and video replay, the evidence was clear and degrading to any Citadel cadet. Right in the middle of the fray was VMI Cadet Carrie McAtee (2008), a cheerleader, with a Citadel rat in a head lock, while all the time she was pounding him in the head. McAtee will go down in Corps lore as one cadet who proved that boxing class has its merits. What other college requires its women to learn how to box?

One afternoon in 2007, while completing my daily walk through town, I watched over several minutes a large number of cadets dressed in ACUs and carrying full back packs (at least 30-35 pounds) run past me one by one or in small groups. After awhile, I thought all the cadets had passed; however, as I entered downtown Lexington, two female cadets came shuffling by bent at 90 degree angles. I could not imagine the torture, they were enduring to qualify and train for commissioning. For that matter, I also wondered what W&L students thought as they paraded past in their J. Crew fashions, flip flops and cell phones. Once again, the VMI females were treated equally and to the standard of their respective military branches. For many females, commissioning is a primary reason for coming to VMI.

While there are few females in our economics and business major, the majority of these cadets have commissioned, and I am sure will make excellent officers. They include Emily Naslund (2005), track team member and Air Force pilot; and Melissa Ward (2003), Tabitha Pinter (2007), and Andrea (Andee) Walton (2008), U.S Marine Corps officers. These women are mentally and physically tough, very determined, and can handle their own weight. Ward has already served in Afghanistan and came back for the 2007 graduation to surprise her sister Abigail, who commissioned Army. The Wards are the first sisters to graduate from VMI.

After briefing Army ROTC cadets on George Marshall's leadership, one cadet stayed behind and asked me what George Marshall would have thought of women in the Corps. I told him all of these VMI female cadets and alumna officers would have made General Marshall proud. Marshall, when as Army chief of staff, was responsible for creating the Women's Army Corps and for appointing its first director, Colonel Oveta C. Hobby. Later while serving as secretary of defense, he named the first woman, Anna Rosenberg, to serve as assistant secretary of defense.

Conclusion

VMI, through use of the Rat Line, Rat Challenge, a mentoring (dyke) and coaching system, and other experiences, has developed a strong systematic program to build a successful and lifetime team of Brother Rats. BRs are educated and trained in the VMI way by a trained cadre of cadets who provide them with feedback and teach them to be followers. Through adverse experiences, cadets discover just how far they can push themselves, and how their BRs need them to succeed as a team or Rat mass. In the process, everyone is treated equally regardless of gender or color of skin. Step by step, cadets take on increasing roles of responsibility and accountability. Through this process of linked experiences, VMI develops leaders of character for our nation.

NEVER SAY DIE
PHYSICAL TRAINING & ATHLETIC DEVELOPMENT

*"The Virginia Military Institute subscribes
to the belief that wholesome and keen competition in
athletics is a vital ingredient in physical and mental
development and that it constitutes a necessary
ingredient to the balanced education of a man [or
woman]."*

1959 *Bomb*

At VMI cadets are taught to believe "Stonewall" Jackson's famous maxim, "You may be what ever you resolve to be."[1] In the fall of 2007 I became a believer. I observed a pugil stick competition between two female Rats. Like other Rats they were squaring off in an attempt to win the match for their squad. The determination to win on these young women's faces was intense. The jabs to the body and head were hard and despite the protective gear worn, I could tell that it was taking all their strength to go the distance. At the end both were exhausted, but they never quit. They had proved that their continuous physical training had toughened them and boosted their self-confidence and self-esteem. Their persistence to the end paid off and won them the respect of their peers and upper-class cadets alike. Their success took hard work, which moved them closer toward their goal of becoming graduates of the Institute.

Some have taken this more seriously than others. Benjamin Franklin Ficklin (1849), described as the "biggest Hell raiser ever at VMI," was expelled from VMI in 1846, when legend has it that he fired an Institute howitzer at the Old Barracks and blew out every window. He is also accused of burying the Superintendent's Francis Smith's boots in the snow, painting his horse with zebra stripes, and detonating fireworks in the Barracks. Ficklin left VMI to join the army and fight in the Mexican War, where he

Female cadets participate in the pugil stick
competition during Rat Challenge.

was badly wounded. He then returned and staged a sit-in on
Smith's office doorsteps. Ficklin demanded a second chance and
readmission to VMI. Smith gave into Ficklin's persistence, and he
finally graduated in 1849. Upon graduating, he stuck his diploma
on a bayonet. He and William W. Finney (1848) later founded
the famous Trans-Continental Pony Express and Stage Line.
Ficklin then served as a lieutenant in the 45th Virginia Infantry,
CSA. [2]

Another distinguished graduate, who classmates
report demonstrated a flair for fun and good times, was Real
Admiral Terrence E. "Terry" McKnight (1978). McKnight
later demonstrated his leadership by serving and commanding
a variety of ships as well as serving as the executive assistant to
the Under Secretary of the Navy. During the invasion of Iraq, he
commanded the U.S.S. Kearsarge (1,200 crew), an amphibious
ship, which transported Marines (1,700), aircraft (24 plus six
Harriers) and equipment to Kuwait. He also served as the
85th Commandant of the Naval District of Washington and
as the Deputy Commander of the Joint Forces Headquarters

National Capital Region. McKnight is currently commanding the Expeditionary Strike Group 2, Task Force 51/59, which in 2009 was conducting a search and capture mission to stop Somalia pirates that are holding ships for ransom off the East African coast. In February 2009, Admiral McKnight was interviewed on national news for his Task Force's successful capture of a boat of pirates.

Bruce E. Baber, author of *No Excuse Leadership: Lessons from the U.S. Army's Elite Rangers* and a former army officer and Ranger, describes persistence:

> *"Persistence does not exist without stress and pressure. You can't persist through a sunny day and an ice cream cone. Your goals require persistence and patience. How many times do you have to get up after falling? One more time. Persistence isn't dictated by beauty and strength. It isn't influenced by a high IQ or perfect eyesight. Persistence is a personal decision made every day or every minute until you have achieved your goal. Persistence is the leveraging of time against the weight of a heavy goal. You'll see that persistence can force actions that are uncomfortable and awkward, and it can compel introspection."* [3]

In the end it will be persistence that wins the War on Global Terrorism. General John Jumper (1966), former Air Force chief of staff, told members of the Institute Society on November 12, 2001, shortly after the attacks on September 11, 2001: "It's not only with the United States Air Force, it's with the coalition, it's with the other services, and we cannot be deterred. There can be no pause until this job is done. However, it's not about one guy. Even when we get Osama bin Laden, it won't be over until we've won the entire war on terrorism."[4] Another VMI graduate, Colonel James Hickey (1982) led the 4th Infantry Division's Raider Brigade that captured Saddam Hussein on December 13, 2003. In an interview with *Pittsburg Tribune-Review* reporter Betsy Hiel, Hickey proclaimed, "We'll get him."[5] Hickey never gave up and planned an aggressive campaign around Tikrit, Saddam's hometown to capture him. Once again persistence and determination won out.

This same persistence is also proven weekly in the football and basketball programs, where for years these teams have not experienced winning seasons. Army General Douglas McArthur once said, "On the fields of friendly strife are sown the seeds that on other fields, on other days, will bear the fruits of victory."[6] Week after week, these dedicated and persistent student-athletes play to win and give their best. Truly this is the spirit of VMI. Not to say that VMI athletes and alumni don't cherish victory, but at VMI as a leader of character, how you win is as important as it is to win. The VMI Athletic Association Constitution and Bylaws (1917) say it best, "Victory is no great matter, and defeat is even less; the essential thing in good sports is the manly striving to excel, and the good feeling it fosters between those who play fair and no excuses when they lose." In 2006, 2007, and 2008, the VMI football team won the Big South Sportsmanship Award.

Athletics and physical fitness are one additional piece of the VMI leadership development system. Together they create cohesion, teamwork, high inner morale, and unit *esprit de corps*. A classic example of athletics and leadership was hatched in the Thunder Dome (Cocke Hall gym), where Isaac Moore 1999 won 30 matches his First Class year and a perfect dual match record of 12-0 on his way to the NCAA championships in wrestling. Second Lieutenant Moore, while serving in Iraq with the 7th Marines, recruited a wrestling team of 18 Marines to take on a group of Iraqi wrestlers from Karbala. The Iraqis were coached by a former world champion, Hussein Khadim. The wrestlers soon met in a cramped 100 degree heat facility; and although the wrestling styles and languages were different, the Marines beat the technique superior Iraqi team in front of a hometown crowd. Moore reports the more physical and aggressive style of the Marines made the difference; and yes Moore won his match. Afterwards Moore arranged for $7,000 to fix and improve the wrestling facility. Thanks to donations, the Iraqi team also has uniforms and shoes for its team and students.[7]

One coach who has been responsible for overseeing physical fitness both on and off the field is Brigadier General (USAR retired) Mike Bozman, a Citadel graduate. Coach Bozman, as most cadets know him, served as commandant of cadets (1994-1996), interim athletic director (1998), and as the head track and field coach since 1985. During his 23-year reign (retired in 2008), he coached 12 conference championships and was recognized as Conference Coach of the Year 13 times. He was also responsible for building the woman's athletic program. Coach Bozman, a Vietnam veteran, served as the platoon leader for a long-range recon platoon. His service in Vietnam earned him a Silver Star, Bronze Star, Air Medal, and Combat Infantryman's Badge. He also earned the Ranger Tab and The Legion of Merit. In 1992, Bozman graduated from the Army War College.

Physical Fitness and Training: Overcoming Fear

Physical fitness can be performance-related or health-related. Performance-related fitness is a measure of an athlete's agility, power, speed, and balance. "The benefits of both types of fitness are a healthy heart (cardio-respiratory fitness) and lungs, increased flexibility, and muscular strength and endurance."[8] The physical fitness program at VMI strives to develop and maintain a vigorous and healthy lifestyle for cadets. To ensure that cadets achieve and maintain a standard of physical strength and endurance, a program of physical fitness is required of every cadet. Education and training for physical fitness is received both in the classroom and in extracurricular activities.

There are additional benefits to physical fitness and training. At VMI, every cadet takes a PE course every semester. Perhaps the one course that produces the most anxiety for cadets is the required Rat boxing course. The purpose of learning to box in some ways has less to do with learning self-defense than it does learning to control fear and aggression. Fear is a basic and unpleasant emotion that is caused by an anticipation of danger or threat whether real or perceived. Fear is not the absence of courage, but rather the courage to act in spite of fear. Well known management guru W. Edwards Deming once said, "Fear takes a horrible toll. Fear is all around, robbing people of their pride, hurting them, robbing them of a chance to contribute to the company. It is unbelievable what happens when you unloose fear."[9]

It takes courage to stand face to face and throw a punch at someone. The art of throwing a jab, a hook, and an undercut are taught as cadets practice these skills by shadow boxing and hitting body bags. Through sheer repetition these acts become ingrained and the fear subdued as the cadet begins to focus on his or her knowledge of boxing skills.[10]

Since most cadets have never been punched in the face, it is very difficult to face this fear. The cadet has two choices, to punch or not to punch. Since they won't graduate if they don't punch, the choice is usually to stand up and take it and learn to dish it out. While some cadets pull their punches praying their fellow cadet will do the same, cadets are taught to hit hard and to win. Some cadets are actually aggressive and seem to enjoy plastering a fellow classmate with a barrage of punches. This now becomes an opportunity to learn self-control on both sides. The first is not to get angry at the person who is clobbering you, and the second is to keep yourself under control should you be winning.

My son, Ryan, shared that his first air assault mission in Iraq was much different than he imagined. His skills as a Black Hawk pilot and his many hours of flying night training missions in Europe and the States took control over any fears that he may have had. He commented that he could have just as easily been flying over Kansas, for the flat terrain seen through night vision goggles looked eerily the same. The anxiety of piloting his first flight in a combat zone left as soon as he was airborne. His self-confidence quickly took over. Part of that self-confidence was learned in a boxing class at West Point, where like VMI, it is a required (males only at USMA) course. Later, it was this same self-confidence that enabled him to serve as the company commander in Iraq of an air assault flight company with the 1st Infantry Combat Aviation Brigade.

The application to the world of business is simple. The global world of business is fraught with chaos and uncertainty. Confusion abounds and employees fear mergers, layoffs, and bankruptcies. Keeping focused on the principles of solid business practices will provide the leader the self-confidence needed to overcome difficult times.[11]

People in organizations have many fears. They include the fear of speaking up, fear of change, fear of the unknown, fear of technology, fear of obsolescence, fear of discrimination, fear of harassment, fear of social sanctions, fear of failure, and even the fear of success.[12] In the mid-1970s, I taught a career development course at Lansing Community College in Michigan. In my class, I had a full-blooded Chippewa Indian, who was bright and on schedule to graduate from our accounting program. With two weeks left in the quarter, he quit school and headed off with a group of his peers to visit some fellow tribesmen in Arizona. His girlfriend told me he would have been the first member of his family and peer group to achieve a college degree. Unfortunately, his buddies were harassing and making fun of him for being a "college man." Apparently, he feared the success his degree would bring and the possible loss of his tribal friendships. He also lacked much of the self-confidence needed to succeed.

Even though "Stonewall" Jackson said, "Never take council of your fears," Jackson feared public speaking. To overcome his fear, Jackson joined the Franklin Literary Society of Lexington. Facing "extreme nervousness and repeated failures," Jackson persisted and forced himself to participate in discussions with Lexington's intellectual elite. He also memorized his lectures and recited them in his parlor to his wife the night before class. He was known to go blank and to start the whole lecture from the beginning rather than start where he left off. Gradually, he became comfortable at speaking and later became a Sunday school teacher and even offered extemporaneous prayer.[13]

The fear of failure and lack of perseverance also rank as top reasons people do not succeed. Author W. A. Clarke once said, "Failure is the line of least persistence."[14] At VMI, the greatest failure is to not try. An effort must be made until persistence eventually wins out. The habit of successfully dealing with thoughts of quitting becomes a part of your thinking process. Of course there will be failures, but cadets are expected to keep making an effort to succeed, if not today then tomorrow. Cadets must learn from their failures by trial and error. Thus, they must be reflective in learning from their varied experiences and analyze their strengths and weaknesses and failures and shortcomings. This self-examination can be painful.

This process also occurs in the formal classroom environment, where in leadership classes, cadets study not just successful leader actions, but also actions that failed. The question is asked, "Why did the leader fail?" As a professor I may or may not be able to teach cadets how to lead, but hopefully I can teach them how to teach themselves how to lead. My goal is to facilitate their ability to relate new knowledge to their experiences whether past, present, or future. For them, the process begins with a battery of self-reflective exercises, a reflective essay, and a journal of leadership observations and lessons learned. They are required to take a hard look at themselves and to set out on a course of improvement.

Doug Crandall, a former West Point professor and editor of *Leadership Lessons from West Point,* believes there are three distinct categories of leadership failure: 1) Level One: Failures in what we do; 2) Level Two: Failures of who we are; 3) Level Three: Failures of who we want to be.[15] Failures at level one reflect on failures at accomplishing specific actions. For example, a cadet may fail to do enough sit-ups or push-ups to pass the VMI Fitness Test (VFT), complete a forced march, or fail to make the right decision in a field exercise. Level two failures are about our emotions, abilities, and personality. A cadet who can't control his or her temper when being boned for a minor offense is experiencing a level two leadership failure. A football player who fails to catch a game-winning touchdown pass may not have the ability to cope with such a stressful situation. At level three, leadership failures are departures from our value systems or deeply held principles. At this level, a cadet could possibly engage in underage drinking or cheat on an exam.

In order to teach cadets about types of failures, I as a leadership instructor, must be willing to examine my failures and offer them as examples. As tough as it is, my life is also my message. Do I lead by example? Do I represent the type of leader that cadets might aspire to be in the future? Hopefully, the answer is yes, but I must always strive to do a better job of demonstrating my leadership skills.

Cadets not on an athletic permit (formal permission to play or practice a varsity sport) have three options to enhance their preparation for the VFT (VMI Fitness Test), which is a required part of every PE course. The VFT score constitutes 25 percent of all PE grades, and thus, cadets are tested each semester. Effective fall of 2008, the VFT is "gender-normed". For a minimum passing score prior to 2008, a cadet had to do five pull-ups, 60 sit-ups, and run 1.5 miles in 12 minutes or less. A failure to meet the standard in any event still results in a zero for that event. While the new gender-normed VFT is certainly controversial, the new standards will be more in line with the armed services testing and the service academies. I feel this change is a step in the right direction. Every cadet (male or female) will have to meet a standard(s) for physical fitness ... not once but every semester.

The PE department is now responsible for measuring the height and weight of cadets to ensure that they do not exceed the authorized body fat percentage. Cadets exceeding the authorized standards must report monthly to the VMI infirmary for appropriate medical screening, nutritional counseling, and body fat assessment. Hopefully, this will put an end to overweight, out-of-shape cadets walking across the stage at graduation. While these cadets were few and far between, there were enough of them (approximately 2-5 per year) that it was clear to me that they were sending the wrong message. The correct message is that VMI's cadets are physically fit and healthy individuals who are ready to answer the call to become citizen-soldiers.

The first option to prepare for the VFT includes participating in a physically demanding club sport such as ice hockey, boxing, water polo, rugby, racquetball, rock climbing, ultimate Frisbee, among many other alternatives. The ultimate

Frisbee team was started in 2006 by two new faculty members from the Department of Economics and Business. Starting with a base of mostly Rats, Majors Sam Allen and Jim Bang built this club sport to a competitive level. The team now participates in a regular schedule of matches and tournaments. Major Allen is now coaching the Marathon Club and can be seen running around Lexington with his runners. Option number two is participating in a voluntary intramural sport. These sports can range from flag football to basketball and vary from semester to semester. The final option is to complete a mandatory workout twice a week.

The workout at a minimum must consist of 100 bent knee sit-ups, 20 pull-ups, and a three mile run. Sit-ups and pull-ups may be done in segments. In 2007, commissioning cadets with their ROTC cadre began doing PT on the Parade Field doing calisthenics and running from 1050 to 1215 on Thursdays. ROTC cadre felt this time was necessary to prepare cadets for the demanding rigors of the wartime environment in which many cadets would soon operate.

Rat Challenge

Rat Challenge is another form of physical training and team building, which is often a highlight of the Rat year. For the most part, Rat Challenge is done away from the public eye, so few faculty or staff have a chance to observe these varied activities. The Rat Challenge course resides behind the barracks and on a ridge the other side of Woods Creek. There, on most Tuesdays and Thursdays in the fall, the majority of the Rat class can be found learning teamwork skills and practicing for the Rat Olympics.

To closely observe this activity, I contacted the senior cadet-in-charge (CIC), Paul Childrey, a mechanical engineering major, and his assistant CIC, Alex Snyder, a chemistry major. The staff and cadets were supervised by Colonel Gordy Calkins, a retired physical education department professor, who volunteered for a year to replace the director. Calkins, a former Marine Corps officer, was all business with the cadets and staff. It was obvious that the cadets respected Calkins and his insights into training methods and operations. He and his staff of other PE faculty, shared the same mission: A dedication to cadet safety and the professional treatment of Rats conducive to teambuilding. Childrey and his cadet staff drove me to the course, where I found a beehive of activity. Cadets were coming and going constantly from each course station.

Cadets Childrey and Snyder explained that the course consisted of a variety of 21 stations ranging from high ropes elements, pugil stick fighting, leadership reaction stations, a Marine obstacle course, the VMI 3.5 mile individual obstacle course, repelling off a 150-foot cliff on the Maury River, five-

Coached by cadre, Rats tackle the wall
with help from their squad members.
Source: VMI Communications & Marketing

mile runs, climbing walls, high-level entry into the Maury River, climbing nearby House Mountain, and of course the infamous Ranger Pit. Each station has 10 goals and objectives. They are to: improve physical condition, help conquer fears, improve relations with cadre, improve self-esteem, help build team spirit, improve ability to resolve conflict, improve sensitivity to differences in ability, develop problem-solving skills, and have fun. Responsibility for each squad of Rats is given to a cadre corporal, who designs his or her own training program. A master schedule of rotations is handled by the CIC and his or her staff.

As someone who operates a private leadership reaction course (LCR), I was amazed in a positive way at what I witnessed. While everyone was 100% business, the atmosphere was much more relaxed than in the barracks. For many Rats, this was their first opportunity to meet their fellow BRs and to have an honest to goodness moment of discussion with them. Cadre, for the most part, were coaching versus "flaming" their charges. They professionally taught their Rats and gave them the necessary feedback to succeed and to avoid being injured. I am sure that everyone was concerned with the safety of all involved and this was obvious. To back up the cadre in case of injury, a VMI EMT is assigned to each station.

The infamous Ranger Pit is the last event
of a grueling day of Rat Challenge.

Other than witnessing the sheer fear generated by rappelling down a 150-foot rock cliff, my favorite event as a spectator was the Ranger Pit. This is the last event scheduled for the day. The reason? The pit, a modified version of the "worm pit" at U.S. Army Ranger School, is actually a five-foot-deep mud hole filled to approximately three feet with water. The cadets jump with mouth pieces into the pit and then proceed to team wrestle six other Rats. The idea is to force the opponent Rats to the side of the pit, where to win, their opponents' chins must touch the top of the embankment. I have rarely seen such a mess. It reminded me of pictures I had seen of the old Breakout ceremony, which was also a mud bath. The only hard and fast rules were no choke holds and no males wrestling females. One thing for sure, the cadre loved this event and cheered wildly for their teams. In the end, the cadets, coated in chocolate-covered mud, were hosed off, and sent to nearby Woods Creek to further rinse their clothes and bodies. These were truly moments of bonding for the BRs.

Athletics: Every Cadet a Student Athlete

Long a member of one of the nation's oldest athletic conferences, the Southern Conference, VMI announced in December 2001 that it would join the Big South Conference in 2003, beginning a new athletic chapter for a new century. VMI presently fields 15 teams at the NCAA Division I level. The sports include baseball, basketball, men's and women's cross-country, football, lacrosse, men's and women's rifle, men's soccer, swimming, men's and women's indoor and outdoor track, and wrestling. The Institute has the third smallest enrollment among NCAA Division I institutions, behind only Centenary and Wofford. Approximately one-third of the Corps of Cadets plays on one of VMI's intercollegiate athletic teams making it one of the most active athletic programs among its student body. Ninety-two percent of VMI athletes, who complete their eligibility, receive their VMI diplomas.

In August, 1997, the first women's athletic program at VMI competed as a cross country team. Since then three additional women's sports have been added, including soccer in 2003. More will likely follow as female enrollment hopefully increases in the coming years.

The earliest sports at VMI were natural spin offs of military instruction. Horseback riding, rifle firing, fencing, ice skating, and hiking in the Blue Ridge Mountains were all popular with cadets. The first organized sport was baseball in 1866. Most early games were against Washington College (W&L), and the first regional game against the University of Georgia was in 1889. By 1880 a gymnastics team was started and in 1885 football was organized as an intramural sport. Colonel Tom Davis, retired VMI history professor, states in his book, *The Corps Roots the Loudest: A History of VMI Athletics*, that the golden age of athletics at VMI was 1890-1920.[16] This era produced many new teams and Corps support that has at times been equaled, but never surpassed according to Davis. From this era emerged the "Sprit of VMI" athletics that spread to generations of cadets and alumni.

Football

The sport of football has a special place at VMI. In January 1930, Colonel R. L. Bates wrote an essay in the *VMI Alumni Review* on football.

"Football is a great character builder. The trend of peace-time civilization is towards the easy living. The trained football player has learned the lesson of restraint. Loyalties and allegiances to one's fellow men and institutions take on a definite significance. Emotions are stabilized. An understanding of the meaning of teamwork is supplied. Like the disciplined soldier, the football player feels the best and safest when his contacts with his associates are the closest, and this is particularly true when the command is under fire. An individualistic spirit of selfness is replaced with a spirit of give and take. To withstand blows without flinching, to hold one's own in the face of adversity and to struggle on undismayed, even when all hope is gone, have distinct parallels in the discipline of life."[17]

Throughout their four years at VMI, cadets are taught to coach others, either through the class system or coaching intramural or club teams. Actually coaching in sports is no different than coaching in management. VMI's most famous athletic coach is 1959 graduate, Bobby Ross. Coach Ross led teams from the Citadel, Maryland, and Georgia Tech to a 94-76-2 record. His Georgia Tech team was the NCAA national co-champion in 1990 and went 11-0 for the season. Ross then moved onto the pros, where he was head coach of the San Diego Chargers and the Detroit Lions. While with the Chargers, the team went to the playoffs three of his five years and to the Super Bowl in 1994 as American Football Conference champions. As a pro head coach, his record was 77-68. After retirement, West Point brought Coach Ross back to the college scene to revive the Army football program.[18] After three exciting seasons, Ross returned to his retirement home in Lexington. He is still seen occasionally on the VMI sidelines mentoring the coaching staff and voluntarily coaching VMI players.

In 1891, the first intercollegiate football team was organized and coached by Cadet Walter H. Taylor III. VMI's first 6-0 victory was against W&L in 1891. The VMI-VPI (Virginia Tech) long-standing series began in 1894. The first game in the long series was played in Staunton and VMI won 10-6. The game then moved to Roanoke in 1896, where the cadets played before 2,000 people. In 1920, the starting team was known as the "Flying Squadron." The team posted VMI's only undefeated, untied record (9-0-0) with defeats over such schools as University of North Carolina, Virginia Tech, University of Virginia, the Citadel, North Carolina State, and the University of Pennsylvania.

The 1920 team consisted of many talented players including Jimmy Leach, who is considered by many as VMI's greatest athlete. Leach is VMI's only representative in the National Football Foundation Hall of Fame. He came to VMI in 1916 and soon joined the Marines and rose to the rank of lieutenant during World War I. After returning to VMI, he commanded C Company, played varsity baseball and served as captain of the basketball team. In 1920, he scored 210 points and 26 touchdowns, both which were national records for many years. It may come as a surprise that in this day of large offensive backs, Leach was a mere five feet eight inches and weighed 155 pounds.[19] Leach was a true consensus All-American.

Colonel R. B. Pogue wrote in the 1921 *Bomb*, "The vital fact in success of the team was the fine spirit of the players who, without exception, were willing to subordinate their own chances for prominence to the general welfare of the teams. We had stars aplenty, but they shone as members of the system rather than individuals."[20]

In 1953, Coach John McKenna became head football coach at VMI. McKenna, a former star lineman at Villanova, was one of only two coaches (Blandy Clarkston being the first), who stayed longer than five years to post more wins than loses. Clarkston and McKenna (1957, 9-0-1 record) coached VMI's only two undefeated seasons. McKenna, however, became a nationally revered and recognized coach. Experiencing losing seasons for his first four years (13-26-1), McKenna went onto to win four Southern Conference Championships and is the only VMI coach to win 18 consecutive games. Coach McKenna left VMI in 1965 to become an assistant coach to the legendary Bobby Dodd at Georgia Tech, where he was later promoted to associate athletic director. McKenna had two other credits to fame. Coach Bobby Ross, who played for McKenna, credits him for his mentorship role in Ross's rise to NCAA (Coach of the Year) and NFL fame. Coach McKenna also coached two Rhodes scholars: Lee Badgett, 1961, (Regimental Commander and second team-All Conference) and Robert C. Randolph (1967).

I will never forget the events leading up to the 1964 VMI-VPI (Tech) game during my Rat year at Tech. Prior to the game, I experienced one of my most memorable hazings. Rats were awakened at 0300 (3 a.m.) and ordered to dress in nothing but our heavy rubber rain coats. We were then sent to join our fellow Rats in the hallways where we bounced up and down like kangaroos until we had to urinate. To force the issue, we were provided gallons of water to drink, while bouncing and shouting "boing, boing, boing." After showering, we were bused to Roanoke's Hotel Roanoke. There we assembled and then marched three miles through downtown to Victory Stadium. In downtown Roanoke, we halted for lunch with family and friends and then proceeded with the parade. Back then, no one could go home for Thanksgiving until the game was over. Fortunately for VPI Rats, the Hokies won 35-13 and life was easier for a brief time after returning to campus after the holiday.

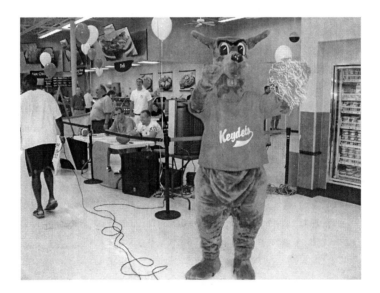

Mo, the Keydet kangaroo mascot, makes an appearance at the Lexington Wal-Mart to promote the upcoming football season. Football and basketball team members enrolled in the author's sports marketing class (summer school 2006) also greeted fans at this co-marketing event.

Probably VMI's most memorable victory in the past 40 years was against VPI in the 1967 Thanksgiving Military Classic of the South. The VMI -VPI Thanksgiving day game was the highlight of every cadet. The excitement building up to the big Turkey Day event was electrifying. I remember that several Hokie cadets rented a plane and dropped flyers over VMI predicating a 68-0 win. Expecting a return raid, the Hokies were on vigil looking for VMI cadets who might kidnap some of our cadets or our beloved cannon, "Skipper." VMI came into the game with a 6-3 record coached by second-year head coach Vito Ragazzo. VMI won the game 12-10. This was an astonishing win, since the year before Tech had defeated VMI 70-12. By now numerous sports writers were predicting this series would not last despite its long tradition and colorful history.

Several members of the late 1960's football teams distinguished themselves as leaders. Brigadier General Robert L. Green (1967) was a member of the 1964 varsity football squad. Green went on to serve in Vietnam as an Army officer and later became the president of Wiley & Wilson, a large architectural and engineering firm. In the mid-1990's, Green returned to VMI and was appointed Deputy Superintendent of Finance in 2000. In 2003, he served as acting Superintendent of VMI before General Peay arrived in July.

During my tenure as a business school dean at another college, Bob Green served as the chairman of our Business Advisory Board from 1994-1995. Green's leadership of our board was a key in the dramatic turnaround of our business school. With his support, we were able to convince college administration and trustees that we needed the funds to build a first-class business school building ($13 million) and excellent business programs, which our region desperately needed. Not to be forgotten, Green's wife, Carol, is vice-president of Alumni Activities for the VMI Alumni Association. She is featured in Laura Brodie's book, *Breaking Out*, for her important role in integrating women into the Corps of Cadets.

Another member of the football team (1968-1972), George "Pete" Ramsey III presently leads the VMI Foundation as its president (appointed in 2004) and is currently serving as a member of the Board of Visitors. Ramsey and I were acquaintances before I came to VMI. Our sons played soccer on the same team, and he also served as a member of our business school's advisory board. Ramsey is president of Taylor-Ramsey, a company that supplies lumber and lumber products worldwide. He is very active in community affairs in Lynchburg including serving as a trustee of Centra Health, the James River Day School, and Junior Achievement. At VMI, he has also served as a member of the Keydet Club's Board of Governors, and the VMI Athletic Council.

VMI won again over Virginia Tech in 1973, 1974, and 1981. According to Donny White, current athletic director,

"The 1973 win (I was the assistant football coach) was 22-21. Eddie Joyce fumbled an attempt to position the ball with a QB sneak on the one or two yard line with time running out, for a field goal. We recovered and ran one play and the game was over. Tech actually tried a field goal about one minute before from the 10 or 12 yard line but they missed it – however, we were off sides and gave Tech a first down inside the 10 yard line. They ran a couple of plays and then tried the QB [to] position the ball – and fumbled. I felt it was justice." [21]

Unfortunately the time-honored series finally ended in 1984.

The 1970's teams, lead by Coach Bob Thalman, were marked by several seasons of winning records followed by losing seasons. In 1977, victories included a 30-6 thumping of UVA and a share of the conference title. In 1981, VMI came back once again to post a 6-3-1 record. The season included an emotional 14-7 win over a highly regarded West Point team. The 1981 season was the last winning season that VMI experienced. By 1984, VMI replaced coach Thalman (30-27-2), but no coach has been able to bring a winning season since 1981. In 2002 and 2003, the team came close under the leadership of Coach Cal McCombs. His teams posted records of 6-6 and 6-6 in these two seasons. At the end of the 2005 season, the popular and personable Coach "Cal," as he was known by his players and fans, was replaced by Coach Jim Reid.

Coach Reid came to VMI as a seasoned and winning coach. He had served as head coach at the University of Massachusetts and the University of Richmond. While Reid did not post a winning season, he brought a new sense of intensity (high energy) and pride to the players and Corps. His players saw him as a "human dynamo" who never stopped working and trying to produce a winning team for VMI. Reid built a team based on the attitude that there should be no excuses. With the completion of the new football facilities, an additional hour of practice, a concentrated effort to red shirt freshman, a renewed emphasis on academics, and a new respect for the Corps, Reid believed VMI football players should "embrace the process" and quit using it as an excuse for not winning. It was his belief that over a seven-year period, VMI could once again produce a winning football team. He told his players that he'd die at the Institute and that he would never quit until VMI won a Big South championship.[22]

Reid started 2006 by insisting his players participate in spring parades and then using the Parade Ground for their agility drills after 1800 (6 PM). He wanted the Corps to see how hard the team worked, and he wanted the team to feel and be a part of the Corps. Coach Reed at 56 years old even went as far as not only observing Rat Challenge, but jumping from a zip line into the Maury River along with the Rats. He liked to see things up front and personally and to feel a part of the VMI experience. This also included such actions as being in the academic halls during summer school to ensure that his players were in class and stopping by faculty offices to introduce himself and his staff. This certainly was impressive to me and other faculty and a sign of commitment to academics as well as football. The Corps also noticed that he showed up at breakfast roll call. He certainly won their respect, and the Class of 2009 made him an honorary member of their class--complete with class ring.

"Embracing the process" became the team motto and was worn on players' and coaches' t-shirts. Coach Reid told alumni, "At VMI you find out quickly there is going to be discomfort sometimes. You have two dramatic choices on how to react. You can whine, complain and cry, you can get yourself in that mode of 'what next?' Or you can say, 'okay, this is what makes us different, so I am going to embrace this for what it is and move on.' That's the way you become stronger. In a game it is the guys who whine and complain who are not successful."[23] Coach Reid demanded total commitment from his team. This was a difficult sacrifice for his players who were trying to balance the Corps, academics, and military duties. Forget a personal life.

According to Reid, "Football allows for opportunities to bring leadership qualities to the surface ... Leadership thrives, we believe, on structure and organization. It's easier for people to lead when they know where they are going, what expectations and goals are, and what their role is within the organization. 'Play it by ear' ... that's not a favorite saying of mine. Put it in writing, organize it and do it --- it's a good way to go".[24]

In January of 2008, the VMI community was very disappointed and surprised to learn that Coach Reid was moving on after two years to become an assistant coach (his "dream job") with the Miami Dolphins. Despite building a foundation for future successful seasons, some cadets and players viewed Reid as a quitter. At VMI, the word "quit" does not rhyme with persistence and determination. It was rumored that some members of the Class of 2009 wanted "their" ring back. They felt Coach Reid broke his promises to the Corps, and thus violated their trust. Regardless of how people feel about Coach Reid, he left a positive mark on the football program. We just wish he could have stayed longer.

Less than two weeks later, VMI announced it had hired Coach Sparky Woods as the new head football coach. Woods is the 30th head coach in the Institute's 117-year history. He brings to VMI 10 years of head coaching experience at Appalachian State University (62-47-5) and the University of South Carolina. Woods was a three-time Southern Conference Coach of the Year

(1985-87) at Appalachian State and Kodak Region II Coach of the Year in 1986.

The 2009 Super Bowl XLIII revealed another coach affiliated with the Institute. The head coach of the Pittsburgh Steelers was none other than 36-year-old Mike Tomlin. Coach Tomlin was the third youngest coach of any major sports league and the youngest Super Bowl winning coach. His career began in 1995 as an assistant coach (wide receivers) under the mentorship of Coach Bill Stewart. In 14 short years, Tomlin was able to rise to a Super Bowl win over the Arizona Cardinals (25-23) and to become the 2008 Motorola NFL Coach of the Year. Even if he is a William & Mary graduate, the Keydets are proud of his accomplishments.

Basketball

The early years (1920-1930) of VMI basketball produced seven winning seasons and two South Atlantic championships. While there were some excellent players and some exciting seasons, nothing of real note occurred before the introduction of scholarships for VMI players in 1958. In 1964, VMI entered the Southern Conference championships with a 9-11 record and was seeded against top 10 Davidson College (21-3). Coached by Wennie Miller, the VMI team played with great tenacity. Led by VMI's only first-team All-American, Bill Blair, the Keydets won 83-80. The Keydets went on to win (61-56) their next tournament game over George Washington. VMI then advanced to the NCAA championships where the team played Princeton, led by All-American, All-NBA, and former U.S. Senator Bill Bradley. The Keydets, played well, but lost 86-60.

Things were quiet for the next 10 years until the 1975-76 season. Led by Coach Bill Blair (member of the 1964 team), who had been at the reins (or perhaps helm) since 1972, VMI teams progressively improved. In 1976, the team finished the season 16-9 with a Southern Conference crown and a birth in the NCAA Eastern Championships. In the first round, VMI shocked highly favored Tennessee by winning 81-75. VMI was now in the top 16 and heading for a second round game with DePaul, coached by veteran Ray Meyer. In overtime, the Keydets won 71-66 in what has been described by some as VMI's greatest athletic victory. The team's next tournament game was against undefeated Rutgers (30-0). Trailing Rutgers most of the game, the Corps chanted "Never Say Die."[25] Despite a hard fought game, the cadets lost 91-75. Coach Blair was name NCAA District 3 Coach of the Year.[26]

VMI continued its winning ways in 1976-77, when the team returned with four starters including Ron Carter. Led by a new coach, Charles Schmaus, the team became the first and last VMI basketball team ranked in the AP Top 20 poll. They were also featured in a February 14 issue of *Sports Illustrated*. The team posted a regular season record of 23-3, including 21 consecutive wins. In the Southern Conference Championship game, they defeated Appalachian State in overtime 69-67. In the NCAA East tournament, VMI defeated Duquesne 73-66, but later lost to a strong Kentucky team 93-78. In 1977-78, with the return of Ron Carter and Dave Montgomery, the team posted a third consecutive (21-7) 20 plus winning season. Carter finished the season with seven VMI records, Southern Conference Player of the Year (twice), and honorable mention All-American. Carter, a battalion commander in the Corps, went onto a career in financial investments. Montgomery earned three school records and fourth place on VMI's all time scoring list. He became an engineer after graduation. With their departure, another VMI era of basketball was over.[27]

In 1994, Coach Bart Bellairs was hired as head coach and in his second season (1995-1996), the team went 18-10. The team produced another winning season (14-13) in 1997-1998. Bellairs was only the third VMI coach to have two winning seasons in the first four years of his tenure. In his rookie year, he was selected by his fellow coaches as Southern Conference Coach of the Year. By 2001, Bellairs became VMI's all-time winning coach. In 2002, VMI defeated Virginia Tech for the first time since 1954.[28]

After several disappointing seasons (2003-2005), Coach Duggar Baucom was hired in 2005 to replace Bellairs who became the Associate Athletic Director for Operations and Marketing Promotions. Bellairs moved on in 2008 to become the athletic director at Savannah State University. Baucom's first season was a tough year for him. He experienced heart problems and was unable to lead his team in his usual enthusiastic manner. The team finished 7-20. The 2006-2007 season was a different story. Baucom came to Lexington ready to play his "run and gun" offense, and this was his season to run. As the season advanced excitement grew among all Keydet fans.

VMI basketball enjoyed banner years in 2006-07 and in 2007-08. Using an up-tempo system for most of the 2006-07 season, the Keydets won the most games (14) in nine years and advanced to the championship game of the Big South Tournament. The Keydets also set new NCAA season marks in 3-pointers, both made and attempted (442 of 1383), 3-pointers made per game (13.4) and total steals (490). Second Class Cadet Reggie Williams became the second VMI basketball player to lead the nation in scoring (28.1). Jason Conley paced the scoring charts as a freshman during the 2001-02 season (29.3 PPG). Williams in 2007-2008 again led the nation with a 27.8 point per game average. In doing so, he became the ninth player to lead the NCAA Division in scoring in multiple years. Sophomore Travis Holmes became the first Keydet basketball player to lead the nation in steals (3.4).[29] His brother, Chavis was fourth in the nation. In late February 2008, both twins were averaging 17.1 points per game.

My first encounter with Travis began in the summer of 2006, when I saw him talking to a friend outside Scott Shipp Hall. I decided to stop and chat about his academic performance and to razz him a little. He stood there never saying a word until I finished, and then he looked at me without cracking a smile and said, "I don't think you know who I am. You must have me confused with my brother. I am Travis." Duh! It was at this point that I realized that Travis had an identical twin named Chavis, who was my student.

The Holmes twins were the second set of identical twins in my classes. Talk about feeling stupid! In their Second Class year, I had both Holmes twins in class. I thought by now I could now tell them apart when they were together, but one or the other would always come by to see me in my office and I'd often get a funny look, when I realized I was messing this up again. One brother would open his eyes as big as saucers and the other one would look at me like I needed a brain transplant. Finally, one day I asked Chavis (or at least I think it was Chavis) how I could tell them apart. Using his dry sense of humor, he said, "I am the good looking one." I guess that is a better answer than my first set of twins, who told me to look at their class rings. You can laugh all you want, but when cadets are dressed in identical uniforms and haircuts, this can become a challenge. Even Coach Baucom admits it took him two years before he figured out, which brother was which. Over their last two years, I developed a positive relationship with Travis and Chavis and always enjoyed watching them play the game.

The Holmes twins, Chavis (left) and Travis (right), enjoy a different game of off court hoops. The twins and Willie Bell, 2008-2009 co-captains, led the Keydet basketball team to VMI's second best season (24-8).

In the 2007 Big South Tournament, the Running Roos were the surprise entry. In the first round, they upset Liberty University 79-78. In the second round, the VMI team once again upset a team they had fallen to twice before in the regular season, as the sixth-seeded Keydets defeated High Point 91-81 to advance to their first Conference Final since the 1987-88 season. VMI played for its first NCAA tournament bid since 1976-77 when it faced Winthrop University at Winthrop. "VMI's storybook run in the Big South Conference Tournament came to an end as Reggie Williams' lengthy fall away three-point attempt failed to drop as the buzzer sounded and Winthrop held on to win, 84-81. The Keydets closed out the season with a 14-19 mark, the most wins for a VMI basketball team in nine years."[30]

The 2007-2008 season ended with a 14-15 record; however, the combined two-year record of 14 wins per season or more was the Keydets' best record since 1977-78 season. Indeed, VMI ranked first again in Division I for scoring offense (91.3), steals per game (12.7), and 3-pointers (11.6).[31]

The 2008-2009 season was one the Keydets will remember for a long time to come. The team, led by senior co-captains Willie Bell and Chavis and Travis Holmes, surprised everyone and completed the season with a 24-8 record including a win over highly regarded Kentucky at Rupp Arena on November 14, 2008. Kentucky head coach Billy Gillispie told reporters that "They whipped us. They were more physical than us. They got all the loose balls...They had a great deal more leadership than we had."[32] Indeed, the three co-captains provided outstanding leadership both on and off the court. Knowing all three young men, I know how hard they worked to transform the players into a competitive team. This leadership included mentoring their Rat team members to insure they survived the Corps side of the VMI experience. When individual players were disciplined with sprints, the co-captains ran with their teammates. Coach Baucom rightfully credits the co-captains' leadership for this season's success.

In the process of producing a team winning season and a second place in the conference, the twins set the NCAA (D-I) record for the most points scored by twin brothers. The record had been held by another VMI pair, Ramon and Damon Williams (3,252 career points). In addition Chavis Holmes finished the 2008-2009 season first in the nation in steals (3.4 per game average) and Travis finished third (2.7). The VMI team finished as the nation's highest scoring team (94 point average) for the third year in a row. Due to the success of the team (second most wins in VMI's history) and their unusual barracks lifestyle, the team became media darlings. Feature articles appeared in *Sports Illustrated*, the *New York Times*, and the *Washington Post*. Jay Leno also mentioned the team on the Tonight Show during his headlines segment. It was a great season for VMI basketball.

Conclusion

A correlation exits between athletics, personal fitness and training, and success in life. Athletics and personal fitness teach us a lot about ourselves. They aid us in discovering or learning where our personal limitations exist. We discover that our bodies need to be fit, if we are to sustain higher performance levels and to remain healthy. Personal fitness sustains our resolve to be all we can be. Both our physical and mental health improves through personal fitness. Athletics teach us the value of teamwork, cohesion, higher internal morale, and *esprit de corps*. It also teaches us to never quit, but rather to fight to the end. That is what the "Spirit of VMI" is all about. In summary, this component of a leader development program teaches future leaders how to lead with determination and persistence.

YOU MAY BE WHATEVER YOU RESOLVE TO BE[1]
LESSONS IN LEADERSHIP

"Few men during their lifetime come anywhere near exhausting the resources dwelling within them. There are deep wells of strength that are never used."

Rear Admiral Richard E. Byrd '08

When describing effective leaders, Dwight D. Eisenhower once commented that leaders must work with and help their followers to do their best. Eisenhower said that forcing people to work was like "bumping up against a brick wall." To illustrate his point, he used a simple piece of string. He'd place the string on the table and say, "Pull the string and it will follow wherever you wish. Push it and it will go nowhere at all. It's just that way when it comes to leading people."[2] And so, VMI must prepare its future leaders to be pullers. Graduates must take the lessons learned from their four years of diverse experiences, critically evaluate themselves, and prepare to motivate and influence followers. Through their example, these leaders will pull their followers to achieve their goals.

In the last nine chapters, we examined the leadership development program at VMI and what makes it unique from West Point and other service academies and universities. You have learned about some standout individuals and their success stories. Understanding that everyone can't directly benefit from VMI's unique program, there are still many "take away" leadership lessons from the VMI experience that are relevant each and every day in organizational settings and situations. To conclude, this last chapter summarizes and provides 25 "take away" leadership lessons learned and applied by many cadets and other successful leaders in their personal lives, communities, and nations. These lessons are offered as a guide to establishing your path to more effective leadership roles. The key to these lessons is ensuring that you follow them consistently. If you fail, your value as a leader may be doubted by members of your organization.

Lesson 1: Know Yourself

What are your strengths, weaknesses, opportunities, and threats? If you can't answer this question, perhaps you should take a battery of self-assessment tests that will help you focus on areas needed for improvement. Conduct your own personal SWOT (strengths, weaknesses, opportunities and threats) analysis. Perhaps you are blessed with people skills and have a gift for communicating with and persuading others. Or maybe you are better at analytical skills or speaking foreign languages, or writing or public speaking skills. All of these are important, but only a small number of individuals can claim all these as strengths. If you understand and know yourself, you can better play to your strengths and avoid your weaknesses until you've had an opportunity to improve them as well. This process will help you discover the leader within yourself. This knowledge will also make you a more valuable team player as each person will likely bring different skill sets to the team.

Lesson 2: Lead a Moral Life

It's easier to pull people along if they observe that you know where you are going and what you represent. Understanding yourself and faithfully following a personal moral compass facilitates the process of leading others. Your character will provide you with the GPS needed to guide you through life. While these words may seem trite and unimportant to some, ask Martha Stewart or other high profile, convicted executives or politicians what lapses in character have meant to their careers and personal lives. One recent example was New York governor Eliot Spitzer, who was known as the "Sheriff of Wall Street" for his criminal prosecutions of Wall Street executives and call-girl rings. Spitzer was forced to resign after he was caught using high-paid prostitutes.

Stand up for your personal values and associate with organizations and people that share similar value systems. Remember that it is important to be ethical all the time and not just when it is convenient. For example, don't talk to your troops or associates about drinking and driving, and then attempt to drive after a "few" drinks at a party. If caught in the military, it will cost you your career. I am aware of one young officer who was serving as a rear detachment commander for his brigade. He was caught DUI. This one act cost a promising officer a position in a highly desired elite unit he was targeted to join and a chance for promotion to major. His family was also hurt. Because he was no longer valued as a trusted and respected leader, he was redeployed back to a staff job with his battalion in Iraq.

Leading a life of integrity and honesty will in turn provide you with the creditability you need as a leader. You will earn your followers' trust if you are creditable. This was one of the major characteristics that led George Marshall to become such a successful leader. He earned the respect of his superiors and followers by being a leader of character.

Lesson 3: Learn from History and Its Leaders

I am constantly amazed at how much we can learn from history. While I am not a history professor, I teach my business leadership course so that it is built around classical characters from history, literature, and film. From these classical characters, we gain a unique perspective of leadership through the ages and throughout a diverse world. Students read the writings of Aristotle, Plato, Machiavelli, Sun-Tzu, Gandhi, Marshall, King, Clausewitz, Lincoln, and others. The purpose of revisiting the thoughts and actions of great historical leaders is that "any contemporary situation is at least partially a product of what has gone on before."[3] Furthermore, history produces long-term effects on trends and influences, which in turn may impact contemporary scenarios, whether economic, political, or social forces, or intellectual developments.[4]

On the flip side, students also learn about contemporary leaders, their challenging situations, and their successful or unsuccessful solutions to problems. Questions are asked such as: What was their decision-making process? What environmental factors, social, cultural, or political, impacted their decisions? How did they motivate their followers? How did history support their decisions? What type of leadership style did they exercise? Were there ethical issues involved in their decisions?

In their book, *Thinking in Time*, Richard Neustadt and Ernest May remind us of how George Marshall used the past as well as the future to make decisions. "By looking back, Marshall looked ahead, identifying what was worthwhile to preserve from the past and to carry it into the future. By looking around, at the present, he identified what could stand in the way, what had the potential to cause undesired changes in direction. Seeing something he had the power to reduce, if not remove, he did so."[5]

Lesson 4: Never Quit Learning

Many college students believe that learning stops when they graduate. Nothing could be further from the truth. Those going off to the military will head immediately to four to 18 months or longer of intense training. Others will go off to graduate school to earn a master's degree or a doctorate, and many more will find themselves locked into a cycle of corporate seminars or professional training required for licensing and renewal of their accounting, physician, dental, and engineering licenses. For those with less time on their hands, on-line learning from thousands of courses is offered by Internet providers including colleges and universities and private and professional organizations such as the American Management Association or the American Marketing Association.

Regardless of your stage in life or education, it is never too late to continue your education and keep on learning something new. In December 2007, I read about an 87-year-old man who had just graduated with a college degree from a state university. As I write this, I also think of my late father. I gave my father a computer (TI-80) for his retirement gift. Some in the family thought he was too old for the latest technology; however, he fooled us all and became an avid computer enthusiast. By the time he passed away, he had upgraded several times and had a better data base of financial and personal records than I ever imagined. He also used a genealogy program to trace family roots and heritage.

Lesson 5: Be Able to Admit Mistakes

The old adage that "you haven't lived if you haven't made mistakes" is certainly true. The real lesson is not whether we make mistakes, but whether we learn from those mistakes or simply repeat them time after time. Admitting mistakes means not just admitting to yourself that you are less than perfect, but also being willing to admit your mistakes to others. Poke some fun at yourself and admit that you were wrong. Having a sense of humor makes this process much easier.

When you are wrong, people expect you to say so. If you don't, then you lose valuable creditability. For example, President Bush told the world that Iraq had weapons of mass destruction. When none were found, the nation expected a quick response from him that he had made a mistake based on faulty intelligence. When he refused to acknowledge a mistake or admit that the war was going badly, he lost his creditability with many citizens of this nation. Critics began to call him a liar. People must trust you if they are to accept you as their leader.

Lesson 6: Put Your Ego Aside

Saint Augustine said, "Humility is the foundation of all other virtues, hence, in the soul in which this virtue does not exist there cannot be any other virtue except in appearance." Humility is a rare quality. Most leaders have an ego. I have found few exceptions. Humility is human nature, but quiet leadership is much more powerful than basking in glory. It also says something about your character. Leaders with big egos are generally not loved.

Their subordinates often follow them for the wrong reasons such as their power or charisma. Thus, the issue is not that you have an ego, but how you display it and how you treat others.

Always treat others with courtesy and give your followers credit for their sacrifices and hard work that made your leadership possible. Never look down on others because you feel superior or make unkind remarks. Show that you are a member of the bigger team and willing to share the limelight. More leaders should lead as Mahatma Gandhi, setting the example with a simple life of humility. Today our business leaders expect to fly first class or in corporate jets, dine at four-star restaurants, and travel in limos. Wouldn't it make more sense not to draw attention, and thus avoid the jealousy and contempt of others?

Lesson 7: Lead with Passion and Energy: Create *Esprit de Corps*.

Leaders are change agents who are responsible for creating a vision and transforming either themselves or their organizations. Your vision for yourself or others provides you with purpose; something to be passionate about and a source of your energy. This energy will radiate through to others, and they will be more likely to want to be a part of what you are proposing. Try to develop *esprit de corps* within your organization. Start customs and unique traditions, which help create morale and spirit among members. For example, ROTC units have annual dining-in ceremonies to teach cadets the customs and traditions of their officer corps. Ceremonies and parades are also held for promotions, awards, and change-of-command. Special uniforms and badges are awarded and worn to create unit pride.

I have a passion for veterans' causes; especially those such as The Fisher House Foundation, which assists wounded veterans and their families. For my retirement, I have a vision of building a retreat center for veterans and their families. A need exists to help vets through the decompression process after combat deployments. With the aid of other veterans and a board of physicians, sociologists and psychologists, and physical therapists, I dream of creating a non-profit to help meet the needs of area veterans and their families.

When I mention my vision to other veterans, they ask, when can we start? What can I do? My vision for these new veterans is shared by many who see the incredible sacrifices they have made and want to help them return to a normal life. Our energy to hopefully make this project a reality will be driven by a passion for helping others, who need assistance.

Lesson 8: Network, Network and Network Some More

Never before have so many people been connected or networked. Our family pays over $300 per month to use technologies like cell phones with text messages, GPS, pictures, and Internet capability; computers with web cams; land-line phones; web pages, faxes; and of course snail mail for packages and Christmas cards to our out-of-town family and friends. Perhaps I should mention that we also use Bluetooth in our car for push button, instant phone calls, which allow everyone in the car to listen as we buzz down the road. I keep thinking this is crazy. Do I really want to be my son's entertainment while he waits at a drive-in window? Of course, the whole purpose of our obsession with communications is to remain connected by establishing and maintaining relationships.

Real connections are about real relationships, not something that happened because we created a page on My Space or Facebook. Sadly, some parents claim they are now registering as "friends" on their child's web site, so that they can track (or stalk) their active social lives. Relationships are about knowing people and caring for them as individuals. Leaders must care for their followers, and there is no better place to start than at home. Whether you are a college student or a business person traveling, you should always make time to touch base with those who care about you.

One evening in the late 1990s I was hosting Jonathan Tisch, CEO of the Lowes Hotel chain. While at dinner, Tisch took a call from one of his young children. He told me this was a nightly ritual when he was on the road. He instructed his kids to call him anytime and anyplace. He believed nothing was more important than staying in touch.

Networking is about using your extended relationships to create opportunities. Always carry a business card, which contains vital information about how you can be reached. Last year, I was asked by a cadet, if I'd help his girlfriend find a job by helping her to improve her resume and cover letter. I did this for her, but I also I did an Internet search for the American Marketing Association in her community. As unbelievable as it may sound, there was a chapter networking meeting downtown that evening, and she was able to attend. Within two weeks she had a new job. Often it is who you know and meet at the right time and place that leads to a job or provides you with valuable business intelligence.

Lesson 9: Keep Yourself Fit and Stress Free

When I first became a business school dean, I attended an AACSB International seminar for new deans at the University of Utah in Salt Lake City. One of the sessions was about maintaining your health and reducing stress. I was amazed that this would be a two-hour topic, and that the problems of deans were so stressful. I soon got the true picture as the session leader asked how many deans had already been sued. To my total amazement, more than half of the 32 deans were already involved in litigation and several had already had heart attacks. I also noticed that more than a few were overweight from the constant banquet food and meals dining out with donors, prospective donors, and faculty and staff recruits.

To help me relieve my stress, I finally purchased a wooded piece of farm land in 2000, and started what I humorously call "chainsaw therapy." Farmers might claim "nothing runs like a Deere," but I'll tell you, I believe nothing runs like a rugged Stihl chainsaw. It can quickly get you back to nature and remove any latent frustration and anger. It doesn't take too many trees to cut and logs to stack to be absolutely exhausted and sore. Of course for others having a hobby or engaging in a recreational activity such as biking, swimming, or hiking/walking might be more helpful in relieving stress.

In all seriousness, regardless of whether you are young or old, it makes good sense to have a physical every year. In addition, you should follow a regular physical fitness program at least four times a week and make an effort to eat less, drink more water, and to eat more nutritious meals. Regular exercise and healthy meals and snacks will keep you in better shape and prevent many illnesses.

Another of my guest speakers in our business school's lecture series was Fred "Chico" Lager, a retired CEO of Ben & Jerry's Ice Cream. Lager was the first CEO of Ben & Jerry's and was very successful at launching the company and giving its eccentric owners the down-to-earth business advice they needed. When we met, Lager was 43 years old and one year into retirement. He told me that he always knew he would quit at 42, because at that age his father died of a heart attack. His father was a workaholic, and Lager was determined not to follow in his footsteps. My time with Lager certainly gave me food for thought and influenced my decision to retire at 62 versus the more traditional 66.

Lesson 10: Anticipate and Respond to Our Fast Changing Environment

Part of teaching strategy to students is teaching them the importance of conducting a situation analysis or SWOT analysis. This includes an environmental scan of your competition, the economy, technology, political-legal, and social and cultural components of our environment. Our environment includes both internal and external forces which can change our organization over time or immediately. For this reason, leaders should become avid readers and observers of trends. For me that includes not only the latest business books, but also reading such publications as *Business Week, The Wall Street Journal, Fortune, Inc,* and others. I also listen to business news networks.

Recently, financial observers learned about the serious effect of subprime mortgages on the housing market and consequently on consumer credit and spending. Observers also noticed how rises and decreases in federal interest rates or initiating federal rebate programs affect consumer spending, which in turn affects retail and wholesale sales, and ultimately the amount of natural resources and energy required to produce products. Leaders must also understand that what may sound like good idea will have a consequence. For example, efforts to use more biofuels as a substitute for gasoline led to fewer cornfields for cattle feed and less for human consumption. Consequently, the prices of corn and beef jumped as more biofuels were produced from corn. Furthermore, the increase in nitrogen-based fertilizers to grow corn has created a large "dead zone" from run-off, where the Mississippi River meets the Gulf of Mexico. Commercial fishing is no longer permitted in this area.

Leaders must also look globally to understand that the impact of what happens in New Delhi or Shanghai will also affect Americans. If American manufacturers sell machinery and robots to the Chinese and Indians, it is likely that their workers over time will earn more and their standards of living will improve. To meet the increased demand for less expensive cars, the Chinese and Indians have started to manufacture their own vehicles and to rely less on U.S. and foreign brands. They also demand more petroleum products, and thus drive up the price of not just gas, but all products made from petroleum derivatives. In late 2008 a U.S. financial crisis, initially caused by a housing-market collapse, created a global domino effect, which additionally included financial industry failures and steep stock declines. Thus, a wise leader must constantly look ahead and prepare for the possible consequences of an ever-changing environment.

Today's leader must also be aware of our changing workforce and consumer demographics. Many companies now market their products in Spanish and target the rapidly growing Hispanic population. How will a shift in labor-force demographics affect the decisions a leader must make? For example, how do leaders treat Hispanics? As citizens or as illegal aliens? Should they ask about immigration status when seeking employees? Is this an ethical or legal problem, or is it just good business to hire the cheapest labor? Some industries like agriculture rely heavily on Hispanic workers. The leader must be prepared to act when faced with these issues, and only a well- prepared individual can make the appropriate choices.

Lesson 11: Take and Accept Responsibility

A real difference between a leader and an ordinary employee is that the leader accepts and often seeks responsibility. Leaders also accept accountability for their actions and those of others. It is their duty. They often perform tasks, which they may not want to do any more than the next person; however, they understand their job is to set the example and to motivate others to accomplish the important tasks at hand. Too many people drift thorough life never accepting responsibility. They are willing to let others fight our wars, volunteer for the local fire department, teach Sunday School, lead their volunteer organizations, coach their kid's Little League team, or whatever avoidance is necessary to never be responsible. A true leader sees what needs to be done and accepts the responsibility of getting the job done. Likewise leaders understand they are accountable for their actions.

Lesson 12: Exercise Self-Discipline

It is so easy to let yourself get out of control. A self-disciplined person can control not only their emotions, but they can say no to themselves, when others may try to push them down the wrong path. Often self-discipline is learned through forced discipline at home, school, and athletics until it becomes a habit. While social networking is a positive use of relationships, it can be a negative if we allow ourselves to fall to peer pressure or to the "if it feels good, do it" moment. The leader remembers for every action there is an eventual consequence. Furthermore, a good leader is an independent thinker, who avoids being unduly influenced by others. Whether it is choosing or declining a second drink or a piece of cheese cake after a heavy meal, life requires self-discipline. It is difficult to follow leaders who can't self-discipline themselves, or who self-indulge and can't say no. Self-discipline is the platform for success.

Lesson 13: Embrace the Core Values of Your Organization

Core values form the foundation upon which we live and work. Before taking a job, define your core values and know what you are looking for in an organization. What is most important to you? Is it your career, family, health, happiness, integrity, wealth, justice, friendship, fame, status, success, or friendships? Hopefully, your personal values will match your organization's core values. If they don't, you and your organization would probably be best served if you looked for another job. One illustration of achieving a match between personal values and organization values is Scott Sayre 1980. Sayre and his wife Mary (his Ring Figure date) started a company out of their garage while Scott was still serving in the army in Germany. Their first product was the very popular elastic boot blouser. The company, now located in Buena Vista, Virginia manufactures a large variety of military accessories, reflective gear, and promotional products. Scott gives every employee a business card printed with company values. Company "important values" include:

- Call us Scott and Mary
- Help others to succeed
- Think like you own this place
- Laugh with each other
- Park further away
- Cleanliness is next to Godliness

The couple considers their employees as family with a culture of performance. The Sayre's run the business so they can maximize their own valued time with each other. The company employs over 100 mostly local residents and is currently the county's largest employer of adults with disabilities. All wage earners receive an hourly wage, and the company aspires to become an employee-owned firm using stock ownership as a performance reward.

It's really tough being in a position, where you are expected to compromise your standards or ethics. Thus, it is very important to select an employer who values integrity, excellence, courage, and customer service. For example, Whole Foods states their core values as satisfying and delighting their customers, supporting team excellence and happiness, creating wealth through profits and growth, caring about their communities and environment, and selling the highest quality products available. These are excellent core values for any organization or leader of character.

Lesson 14: Demonstrate Care and Respect for Others

Respect the dignity of everyone you come across, whether fellow employees, customers, acquaintances, or strangers. Many people come to work with heavy burdens that can potentially affect their work performance. In my work environment, I have experienced individuals with terminal cancer and other serious illnesses; children or spouses at war or with life threatening illnesses; and those who have lost homes in fires, were injured in auto accidents, or were facing divorce. These individuals were hurting inside and needed compassion and understanding from those around them.

Each of us is surrounded by those with difficult situations. If we expect to pull these people towards our goals, we need to care deeply about every one of them. At times this may be tough. People who are experiencing difficult times can be distant, emotional, or even cranky. In our own personal ways, we need to reach out to them with a card, a phone call, or a personal visit to let then know that we care and are there to help if needed. Sometimes you can help by offering them a flextime schedule or an extra day off. My wife, who is also a college professor at another college, was offered an entire semester off, when it was discovered she had breast cancer. What amazed us was that she didn't even have to ask. That deep sense of caring made a very positive impression on both of us.

Lesson 15: Be Loyal

Loyalty is less about what others expect from you in terms of devotion and commitment to them, than it is about your commitment to your own principles and values. Many leaders from kings and presidents to generals and CEOs have demanded loyalty from their subordinates; however, loyalty should never be blind. Leaders who demand loyalty (versus earn it) do not want their followers to betray them or talk about them behind their backs. Some followers have remained loyal to the point of going to jail for not betraying their employers. One would assume they shared the value of loyalty to an extreme. When choosing loyalty, make sure the cause, principle, or person is worth the value of your loyalty.

As a dean, I tried to use a participative leadership style. I would meet with my department coordinators, and we would hash out issues. While it was alright to disagree behind closed doors and to vet concerns, once we decided on a course of action that was it. My goal was to try to achieve 100 percent buy-in, but I learned this was not always possible. Regardless, once made, I expected loyalty and support for our decisions. Unfortunately, some members of the leadership team didn't think loyalty was important. They went back to their departments and told others they were not in favor of decisions made and then they proceeded to undermine our work as a team.

I observed the same thing occurring at a higher level on college-wide committees. For example, the business school personnel committee and I might recommend tenure or promotion for a member of the school faculty. A business faculty member serving on the college personnel committee might then lobby against tenure for personal reasons and vote against his or her own school's recommendation. These faculty members served on the college personnel committee to represent and communicate the department's position on personnel matters ... not the other way around.

Staying on message is very important once meeting doors are open. I expect team members to be advocates and to support the decisions I have to make when serving in a leadership position. When private discussions start occurring by disgruntled team members in the classroom or in open halls, then I believe these individuals should be counseled and told these conversations are inappropriate. If that doesn't change their behavior, termination is the next step. For those readers familiar with academics, you know that firing a tenured faculty member is nearly impossible; however, a documented record of insubordination and counseling is always the preferred choice for terminating anyone's employment. Failure to perform as told is acceptable in situations involving ethics or unlawful practices. In work settings, a minority should never rule when the majority has voiced an opposing opinion in a fair vote.

Lesson 16: Be a Servant Leader

Acts of service are extended acts of care and concern for others. Selfless service often begins with small acts of helping others and later grows to major commitments of time and money. Stated another way, selfless service is about putting others before your personal ambitions and goals.

In many small communities, business owners and their employees can be seen leaving work to assist in putting out fires or rescuing people in the middle of the day or night. For example when visiting my in-laws in northern Michigan, I heard the fire siren go off one afternoon. I was standing in the checkout line at the local IGA grocery store. The grocery owner, who was running the cash register, said "sorry I've got to go," and he ran out the door for the fire station. Later in the afternoon, I saw him manning a bulldozer to create a firebreak necessary to stop a forest fire from spreading too close to town.

Others serve their nation in the National Guard or their communities by volunteering at the local library, hospital, soup kitchen, or raising money for local charities. Selfless service is at the very core of our self-being. We exist on this earth to serve and help others who may not be as fortunate as we are. Good leaders ensure that they set a positive example for all.

Lesson 17: Demonstrate Personal Courage

In the military, we most often think of personal courage as bravery displayed on the battlefield. However, most of us will never experience the need for that kind of courage. Rather than address acts of bravery (ultimate courage), we should examine the acts of personal courage that face us in our daily lives. President John F. Kennedy once wrote, "In whatever arena of life one may meet the challenge of courage, whatever sacrifices he faces if he follows his …conscience – the loss of friends, his fortune, his contentment, even the esteem of his fellow men – each man must decide for himself the course he will follow. The stories of past courage can define that ingredient – they can teach, they can offer hope, they can provide inspiration. But they cannot supply courage itself. For this, each man must look into his own soul."[6]

Caroline Kennedy, in a follow-up book (*Profiles in Courage for Our Time*) to her father's best seller (*Profiles in Courage)*, writes of courage in the pursuit of justice and the journey of conscience by a diverse group of public servants. Kennedy speaks of the courage required of politicians to cross political lines to vote their conscience and to compromise and build consensus; the courage of those who stay the course even when unpopular; and the courage of politicians to reject pressures of special interest groups. Leaders must confront problems directly and ethically. As future or present business, military, government, or nonprofit leaders, you must stand up and be counted when you see an injustice in the making or an unethical practice being conceived.

Kennedy cites Senators John McCain and Russell Feingold, representing both the Republican and Democratic parties, who created a partnership to co-sponsor legislation (2002 Bipartisan Campaign Reform Act) to limit Congressional candidate spending and to ban soft money (unregulated and unlimited donations by special interest groups and wealthy individuals). During races to retain their seats, special interest groups representing both parties launched negative campaigns to aid McCain and Feingold opponents. Both senators survived their races and Senator McCain ran twice for the presidency of the United States. In 2007, in a speech at VMI McCain said he would rather lose his bid for president than lose the war. His position at the time was very unpopular with the media and public, and like his previous position on soft money and campaign restrictions, it required a lot of personal courage.

Lesson 18: Strive for Excellence in All You Do

Day after day, I witness people who give less than what is required to achieve excellence in their personal lives and in their jobs. At colleges, some students don't take notes or read their texts (assuming they purchased one), and many often wait to the night before to write a paper or study for a test. Consequently, I might later hear the same students rant about how much time they spent working on a project or studying for a test; they can't understand why their grades are lower than they expected. Excellence requires effort, but it is not about how much time you spend on a project or studying. It is all about the quality of what you produce as a good or service. Time simply does not equate to quality. Excellence means attention to detail and effort to do the best possible job we can. It means going the extra mile. Every person and organization should strive for excellence. As the Army slogan says, "Be all you can be." Anything else is cheating yourself or your organization.

Lesson 19: Be Able to Make Your Own Decisions

Leaders are paid to make decisions. Too many people rely on others to make their decisions for them. Perhaps part of this is due to a "helicopter generation" of parents, who are always there to pick up the pieces for their children. They "hover" over their children waiting for them to make a mistake. This often makes a child an emotional cripple, afraid that if they make a decision, it will be the wrong one. Rather than make a decision, they procrastinate and let someone else step in when things start going wrong. That way, other people can be blamed for their failures.

In today's fast-paced environment, decision makers must be able to make decisions quickly and decisively. In the military, leaders learn that fast is not always perfect, but a lack of a decision can cause chaos and crisis situations. The best way to make a quick decision is to quickly cut through the chaff and boil complex facts down to the essence of what is important. The Marines live by a 70 percent solution, which dictates making decisions based on goals and plans with a reasonable chance of working.[7] If 70 percent of the facts are available, then it is time to make the decision. You cannot wait for perfect conditions or all the facts. They may never come.

Lesson 20: Demonstrate Confidence In Yourself

Not too long ago I had a VMI cadet assist JROTC cadets at the Goose Creek Leadership Reaction Course (LDC). If I had predicted which of my nine participating cadets would mess up, this young man would have been at the top of my list. Each VMI cadet was evaluated on his or her leadership skills as they facilitated the day's activities. An independent evaluator conducted the assessment. At the end of the day, the evaluator told me which cadet had stood out as a leader. I had also observed the same leadership behaviors, which made me feel the cadet understood the classroom theories and the essence of what I as trying to teach in the classroom. In the next class, I singled the cadet out for praise and told his peers what an outstanding job he had done; and I cited examples of his performance.

Once again risking failure and mistakes and accepting responsibility are necessary to give people the self-confidence needed to succeed. Knowing you can succeed builds self-confidence. Confidence breeds good leadership.

Lesson 21: Be Determined to Achieve Your Goals and Demonstrate Perseverance

Sometimes I ask myself, is there is a difference between stubbornness and the perseverance to see things through?

Presently, this question is frequently discussed in the news media. Was President George W. Bush stubborn or merely showing perseverance with his decision "to stay the course" in Iraq? If our goals are to achieve victory in Iraq and attempt to stabilize the Middle East, then this position might be viewed by some as hanging in there and doing what is necessary to defeat terrorism and Muslim extremism. To these individuals, this is perseverance. However, others view Bush's decisions as acts of stubbornness, failing to admit that mistakes were made at the beginning of the invasion, and then pushing on to prove he was right in the long run, a sort of, "the end justifies the means" approach.

Being determined to achieve your goals and demonstrating the necessary perseverance to accomplish them is an admirable quality. However, we also must be willing to look back over our shoulder and see if our goals are still valid and our strategies still achievable. A good leader knows how to scan the operating environment and determine if assumptions made and plans created are still adequate for the mission. A good leader adjusts his or her goals and plans and is flexible when necessary.

Lesson 22: Look and Be Professional

First impressions are everything. Many people would debate the meaning of looking professional. So what does looking professional mean? To most business school professors, looking professional means that men wear a coat and tie and women should wear a business suit or appropriate skirt and blouse. The funny thing is, we as professors, are a bit hypocritical. As a dean I had guys who wore "dress" t-shirts, bib overhauls, jeans, polo shirts, and sandals. They drew the pants line at shorts; however, when I was a graduate student at the University of Hawaii in the early 1970's, some professors did wear shorts, Aloha shirts, and sandals. I've also seen women at work in outfits I am sure I don't want to describe, but I think you get the picture.

When discussing this issue with our cadets, they tell us that many of the businesses where they work in the summer have casual days and other cadets tell us they wear uniforms (delivery services, hospitals, security services, and hotel and travel industry jobs to name a few). At VMI the faculty and staff wear uniforms. A few years ago, a professor told me that a student in his class refused to wear a sport coat and tie (standard attire for business presentations) for his group presentation. The professor marked him down and the student filed a complaint against the professor. The student claimed he didn't own a sports coat and to ask to borrow one would be embarrassing and demeaning.

The truth is many students don't have a coat and tie at college, and it has little to do with money. They choose not to bring or buy one, because they have little need to wear one. However, if I am going to educate people to work in an office setting as a leader or future leader, then they need to understand that the uniform of the day is a suit or sports coat and tie in most corporate settings. For women it is a business suit or other appropriate attire, which fits the situation. Smart graduates will take some of their graduation gift money and buy suitable business attire. A nice, conservative suit (dark blue or gray with or without pin stripes) never hurt anyone, and it still makes a positive impression. Dressing down never gives the same impression. Believe me, it is much easier to take off the tie and jacket and roll up the sleeves if necessary, than to show for an appointment looking like you just rolled out of bed. Need I say you also need to look well groomed … have a haircut, fresh shave, clean nails and polished shoes.

Finally, being professional is more than appearance. It is also an attitude which includes being punctual, respectful, courteous, attentive, helpful, loyal to organizational values, ethical, and other attributes, which sound like the Boy Scout laws.

Lesson 23: Find a Mentor and Become a Mentor

I have had several mentors during my career, but none as close as my first department chair at Indiana State University. We are still in touch with each other by email or phone almost weekly. He is now in his late 80s, but I still value his wisdom and the manner in which he mentored me. He always encouraged and supported my ideas, and when I made weak decisions, he would coach me. He never raised his voice to me, always had a smile, and was interested in what I had to say. Dr. Ralph Mason led by example.

Now that I am older, I have tried to find young faculty who may need assistance or guidance. I try to informally mentor them, but I never force myself on them. I know I am doing my job when they seek me for advice. At VMI we also use a formal mentoring program, where every new faculty member in our department is assigned an experienced mentor. I also try to do the same thing, when I see students who need that extra encouragement. Students especially need mentors, whom they can call for advice about job situations, graduate school, or other career related matters, even after they graduate. This may or may not be their advisor.

Lesson 24: Be a Problem Solver

Problems in life are like the farmer who said, "The hardest thing about cows is they never stay milked."[8] Problems will always be with us, but the wise leader is one who senses problems before they occur; thus, avoiding a waste of valuable time. Good leaders are seldom blind-sided when people bring problems to them. Never solve a problem for a subordinate, rather solve it with them. It is extremely important that you develop your subordinates, so in the future they can resolve similar situations. Keep problem solving at the lowest level possible.

As we all know there are some really smart people out there. Unfortunately, some of them are not blessed with the ability to solve the most practical of problems (fix a flat tire, repair a leaky faucet, organize a dinner for friends, relate to children and spouses, plan a fundraising event, solve social or interpersonal problems, etc.); however, give them something technical, mathematical, computer related, or a scientific problem within their area of interest, and they are pure geniuses. As you climb the leadership ladder of success, you will have fewer problems, but much more important ones to solve. To solve today's most complex problems, you will also need a basic knowledge of technology, engineering, science, mathematics and computers, but you will still need to demonstrate good practical sense.

To be effective in a business or other organizational structures, you don't need to be an expert in every field, but you do need to know enough to recognize what problems you can tackle, and which you should pass off to someone better qualified in the organization. Don't be afraid to ask questions of your subordinates, gather that information, and check their sources and resources. Problems, which you should take on include emergencies, those involving organizational leadership, those requiring your unique skills, knowledge, or experience, and those where your followers are stuck.[9]

Don't be afraid to seek solutions, which may take you out of your comfort zone or your traditional ways of doing things. You need to be creative and seek innovative solutions to problems. This is the way societies and organizations advance. Use your imagination in brainstorming sessions as one method to discover innovative solutions and then always analyze a set of alternative solutions. Our nation's future is as good as our ability to adapt to our environment and to solve problems with creative solutions. Creative leaders are needed to meet the needs of a growing global population.

Lesson 25: Learn to be a Valuable Team Member by Motivating Others

Most leadership books talk about building the team. While carefully selecting and hiring team members is certainly important, it is equally important to be a valuable team member whether you are the leader or a follower within the team. Your most valuable contribution may be to have in place procedures and processes for handling problems before they are needed. Also don't be afraid of bringing in others regardless of their background, rank, or position in the organization, if they can assist in solving the problem.

As a team member you are in the position to motivate others and to create a sense of urgency. To avoid panicking your team, assist them in setting priorities. Whenever we do team projects in class, I notice that many teams have a hard time focusing and deciding direction. Since team members aren't always best friends and may not know each other, your role on the team might be to start building rapport and developing stronger relationships between members and with you. If the team is larger, it is essential that this be done one person at a time. Encourage interaction between you and the team and try to influence them to focus on the organization's goals. Socialize with them and over time you will see your team as a second family. Determine some unique team rituals such as meet with them every Monday morning for coffee at Starbucks, have them for dinner once a month, participate as a team in basketball, softball, or ultimate Frisbee.

Conclusion

Being a leader of character is not always the easiest thing to do on a consistent basis; however, most of us want to work for those leaders, because we know we can trust them as individuals of integrity. Most will agree that leaders are made and not born. Each of us has latent leadership potential, which we must develop if we are going to achieve our best efforts and results. It is rarely too late to get involved in your organization's leader development program. In reaching this point in your career (especially in early or mid-term careers), make sure you move yourself through varied and increasingly responsible management experiences. Experience will give you the skills and self-confidence necessary to handle more complex problems. It will also give others more confidence in your abilities and gain you their support. Later, they could be the key to successful changes needed as you develop your vision, plans, and execution for new ideas, products, services, programs, and even new organizations.

GLOSSARY OF VMI TERMS

Academic Rat Line. A Rat's academic course load, which can be extremely tough and hard on the GPA.

Academic Time. Time set aside in a cadet's day, which must be devoted to studying.

ACUs. Army combat uniform, which replaced the battle dress (BDUs) uniform in 2006.

Battle of New Market. The Civil War battle on May 15, 1864, at New Market, Virginia where 241 members of the VMI Corps of Cadets victoriously engaged Union troops.

Blue Book. The cadet handbook of regulations and rules of daily living.

Bone. The act of giving demerits.

Breakfast Roll Call (BRC). The formation for breakfast.

Breakout. The annual event in late January, when Rats are put through various physical workouts and ceremonies to finally earn the privilege of becoming Fourth Class cadets.

Brother Rat (BR). A Rat's classmates who matriculated to VMI in the same year.

Cadet Sergeant of the Guard. Second Class cadet who works with the Cadet Officer of the Day to command the daily Cadet Guard team.

Cadet Officer in Charge (COIC). The cadet responsible and in charge of an event.

Cadet Officer of the Day (OD). First Class cadet in command for 24 hours of the daily cadet Guard team. Only cadet allowed to bone cadets for violating rules.

Cadre. The cadet team of officers and other personnel responsible for training Rats and those chosen to lead the Corps at various levels.

Class System. A system of regulations and privileges which guide each class.

Code of Honor. Each cadet signs a pledge when they matriculate, which states they will not lie, cheat, steal, nor tolerate those who do.

Commandant of Cadets. An administrator who is the equivalent to a dean of students at civilian colleges and universities.

Commandant's Time. Time set aside in the daily schedule to accommodate inspections, parades, marches, and other Corps related activities.

Corps of Cadets. The unit of young men and women enrolled at VMI.

Cross-dyking. Reference to female or male First Class cadets being matched as dykes with a Rat of the opposite gender (no longer permitted).

Dean's Time. Time designated for use by the dean or faculty for speakers, additional classes, test or exam make ups and other academic purposes as designated by the dean or academic chairs.

Demerits. Penalties assigned by the commandant's staff for violations of the Blue Book.

Dinner Roll Call (DRC). Lunch formation.

Drum Out. Event occurring in barracks after a cadet is found guilty of an honor violation and consequently is dismissed from the Institution. The event begins with a long roll of the drums.

Dyke. A First Class cadet who serves as a mentor to a Rat. Term also used to refer to the Rat in the relationship.

Executive Committee. Subcommittee of the General Committee. Responsible for serious cases of misconduct and discipline in the Corps.

Field Training Exercise (FTX). A military exercise often lasting three to four days.

First Class Cadet. A senior cadet.

Flamers. Cadets who make it a habit to scream at Rats and make them suffer.

Fourth Class Cadet. Freshman cadets who are no longer Rats and have endured Breakout.

General Committee (GC). Team of cadets composed of class officers and the president of the OGA. RDC, OGA, and the EC are subcommittees of the GC. The GC is responsible for the appearance and standards of the Corps.

Guard Duty. Twenty-four hour duty assigned to a cadet company on a rotating basis of every nine days. Cadets on guard march in front of the barracks and serve a variety of other duties depending on their class.

Gym Dyke. Cadet uniform for physical training and other physical activities.

Hay. Cadet mattress and bed, which must be stacked and rolled daily.

Hell Night. The night Rats meet the Corps at a sweat party.

Honor Court. Nine member cadet court, which determines innocence or guilt of cadets accused of violating the honor code.

Institute Business. Language used by cadets and faculty to describe their involvement in an honor investigation or trial.

Junior Reserve Officer Training Corps (JROTC). Character-building program for secondary students.

Leadership Reaction Course (LRC). An outdoor course designed to challenge the physical and mental abilities of an individual through a variety of problems.

Matriculate. A cadet who enrolls at VMI; a cadet matriculates to VMI on Matriculation Day.

Matthew, Luke, Mark, and John. The names for the four Civil War cannons in front of the barracks.

Minks. Nickname for Washington & Lee students.

New Barracks. The middle barracks built in 1949.

Penalty Tours (PTs). Punishments assigned for excess demerits and marched in one-hour periods of time.

Permit. Administrative paperwork required to go off Post for various activities.

Permit worms. Cadets who frequently request permits to participate in sports or other extracurricular activities, which excuse them from military duties.

Officer of the Guard Association (OGA). A group of First Class cadets responsible for the cadet morale; reports to the First Class president.

Old Barracks. Completed in 1851 and survived bombardment by General Hunter's artillery.

Old Corps. Reference by cadets and alumni to the way it used to be at VMI.

On the Gim. Sick call list. Cadets routinely on the list are said to be "riding the gim."

Ranger Pit. Infamous mud hole and last event of Rat Challenge.

Rat. A freshman cadet until Breakout.

Rat Bible. Pamphlet of rules and regulations unique to Rats, which must be memorized.

Rat Challenge. A series of physical competitive events, designed to build Rats into high performing teams.

Rat Crucible Day. First full day of training by cadet cadre of the Rat mass.

Rat Daddy. Upper-class cadets who treat Rats kindly and protect them from flamers.

Ratline. The entire scope of the Rat disciplinary system, but refers technically to the lines inside the barracks, from which Rats cannot stray.

Rat Disciplinary Committee (RDC). A subcommittee of the GC composed of First Class cadets, which disciplines Rats for violations of Rat Line.

Rat Mass. The collective group of Rats, who have not earned the right to be called a class.

Regimental System. The military system, which governs cadet life.

Resurrection Week. The period of time before Breakout.

Ring Figure Weekend. Social weekend when the Second Class receives their rings and forms their class number at the dance. First held in 1926.

Reserve Officer Training Corps (ROTC). Senior military programs of the Army, Air Force and Navy designed to commission cadets as officers. First established at VMI in 1919.

Run the Block. To leave Post without a permit or permission.

Second Class Cadet. A junior cadet.

Smack Rat. Rats who are not serious about the Rat Line.

Stoop. Balconies ringing the inside barracks perimeter.

Strain Night. The first night after parents leave and the cadet cadre begin the education process of becoming a cadet. Rats are required to suck in their guts and chins and assume the position of extreme attention.

Summer Undergraduate Research Institute (SURI). Research program of study between a faculty mentor and a cadet conducted during summer terms.

Supper Roll Call (SRC). Closing formation of the day for supper.

Straining. The sucking in of gut and chin by Rats in a stark form of attention.

The Supt (Superintendent). The equivalent of a college president; the ranking general on Post.

The Institute. The Virginia Military Institute.

The VMI Spirit. Song written in 1916 by Benjamin Bowering, which is sung at athletic events.

Third Class Cadet. A sophomore cadet.

Virginia Mourning Her Dead. Statue sculpted by Moses Ezekiel '(1866).

VMI Doxology. Spirit song sung to tune of Christian Doxology.

Undergraduate Research Initiative (URI). A major cadet research project conducted during the academic year under the supervision of a faculty member.

NOTES

Chapter 1 Notes

1. J. Thomas Wren. *The Leader's Companion: Insights on Leadership through the Ages* (New York: Free Press, 1995); and George Manning and Kent Curtis, *The Art of Leadership* (New York: McGraw-Hill, 2007).
2. Bill George. "Truly Authentic Leadership," *U.S. News & World Report*, October 30, 2006, 52.
3. Bill George. "America's Best Leaders," *U.S. News & World Report*, October 30, 2006, 52-94.
4. Ibid, 56.
5. George, "Truly Authentic Leadership," 52.
6. Stewart MacInnis. "O'Connor Praises Cadets for Their Commitment," *The Institute Report*, April 2008, 2.
7. James Kouzes and Barry Posner. *Leadership Challenge*, 3rd ed., (San Francisco: Jossey-Bass, 2002).
8. Bill George and Peter Sims. *True North: Discover Your Authentic Leadership*, (San Francisco: Jossey-Bass, 2007).
9. Ibid.

Chapter 2 Notes

1. Rick Atkinson speech at VMI on February 26, 2007.
2. Ibid.
3. Ibid.
4. Henry Wise. *Drawing Out the Man: The VMI Story*, (Charlottesville: UVA Press, 1978), 29.
5. "Author Stresses to Cadets Importance of Studying History," VMI web site www.vmi.edu, February 25, 2008.
6. Harry F. Byrd, Jr. Speech delivered at a 1984 Virginia meeting of the Newcomen Society of the United States held in Richmond, Virginia, March 26th, 1984.
7. J.H. Binford Peay III, "Remarks to the VMI Faculty and Staff," *VMI Alumni Review*, Summer 2003, 3

8. Henslin, *Sociology: A Down-to-Earth Approach*, 7th ed., (Needham Heights: Allyn & Bacon, 2005).

9. Len Marrella. *In Search Ethics*, (Sanford: DC Press, 2001), 173-174.

10. William A. Cohen. *The New Art of the* Leader, (Paramus: Prentice-Hall, 2000), 65-72.

11. J.H. Binford Peay III, "The Virginia Military Institute: Educating Leaders and Citizen-Soldiers for an Uncertain Future," *Richmond Times Dispatch*, August 21, 2005.

12. Stewart W. Husted and Clifford West. "Developing Leaders: Does a Military Education Make a Difference in Today's Business World?," 2003.

13. J. M. Kouzes and B. Z. Posner. *Leadership Challenge*, (San Francisco: Jossey-Bass, 2002).

14. Laurie F. Brodie. *Breakout*, (New York: Pantheon Books, 2000), 7-8.

15. Stewart W. Husted. *George C. Marshall: Rubrics of Leadership*, (Carlisle: Army War College Foundation, 2006), 9.

16. Josiah Bunting III. Speech to 2002 Institute Society dinner and reprinted in the *VMI Alumni Review*, Winter 2003.

17. Bob Holland. "Triplets Survive Rat Line," *The Institute Report*, April 2008, 1.

18. Doug Cranwell. *Leadership Lessons from West Point*, (New York: John Wiley & Sons, 2007), 69.

Chapter 3 Notes

1. "Learning to Lead," www.learn-to-lead.org, September 2008.

2. George Manning and Kent Curtis. *The Art of Leadership*, (New York: McGraw-Hill, 2005), 3.

3. Ibid.

4. Ibid.

5. Mike M. Strickler. "The Reverend John J. Jordan, '52 Speaks to the Corps," *VMI Alumni Review*, Spring, 2002, 6.

6. Manning and Curtis, 3.

7. Bill George. "America's Best Leaders," *U.S. News & World Report*, October 30, 2006, p.52.

8. This model is not official and represents the observations of the author and interviews with BG Casey Brower, former VMI Dean and Acting Director of the New Leadership and Ethics Center.
9. Laura F. Brodie. *Breaking Out,* (New York: Pantheon Books, 2000), 25.
10. Ibid.
11. "Bobby Jones: Stroke of Genius," www.PluggedIn.com.

Chapter 4 Notes

1. James M. Kounzes and James Z. Posner. *Credibility.* (San Francisco: Jossey-Bass, 1993).
2. Thomas J. Neff and James M. Citrin. *Lessons from the Top.* (New York: Currency Doubleday, 1999), 380-387.
3. Albert Z. Conner. "Moses Ezekiel," www.jewishvirtuallibrary.org.
4. Julian E. Barnes. "An Open Mind for a New Army," *U.S. News & World Report,* October 31, 2005, 72-73.
5. Alan Axelrod. *Patton on Leadership,* (Paramus: Prentice Hall, 1999), 203.
6. Ibid.
7. Henry A. Wise. *Drawing Out the Man: The VMI Story.* (Charlottesville, UVA Press, 1978), 28.
8. Ibid, 37.
9. David H. Freedman. *Corps Business.* (New York: HarperBusiness, 2000), 147-148.
10. Jim Richardson, *The Spirit,* (Louisville: Harmony House Publishers, 1994), 73.
11. Melissa Burden. "Fallen Soldier Lived to Lead: Flint Military Family Mourns Beloved Son," *Flint Journal,* September 4, 2005.
12. Calvin R. Trice. *Richmond-Times Dispatch,* June 4, 2004.
13. Sherri Tombarge, "Alumnus Works to Mitigate Scourge of HIV/AIDS in Africa," www.vmi.edu, July 9, 2008.
14. Rich Meredith, email to ECBU faculty and staff, 2007.
15. Charles W. Eagles. *Outside Agitator: Jon Daniels and the Civil Rights Movement in Alabama,* (Chapel Hill: University of North Carolina Press, 1993), 15-17.
16. Ibid.

17. Kimberly Hefling. *The Lexington Herald-Leader*, June 22, 2004.

18. Richard E. Byrd. *Alone: The Greatest Antarctic Adventure of Our Time*, Island Press (paperback), 2003.

Chapter 5 Notes

1. G. K. Chesterfield. *Letters*, 374.

2. Larry R. Donnathorne. *The West Point Way of Leadership*, (New York: Currency/Doubleday, 1993), 67-69.

3. Warren Bennis. "The Leadership Advantage," *Leader to Leader*. Leader to Leader Institute, Spring, 2003.

4. Stewart Husted and Cliff West. VMI Executive Alumni Survey, 2003.

5. Ibid.

6. Ibid.

7. James D. Hunter. *The Death of Character*, (New York: Basic Books, 2000).

8. Les T. Csorba. *Trust: The One Thing That Makes or Breaks a Leader*. (Nashville: Thomas Nelson, 2000), 65.

9. Stewart W. Husted. "The National D-Day Memorial: The Right Vision Becomes a Question of Fundraising Ethics," *Business Case Journal*, Winter 2007.

10. *Alumni Review*, Spring 2003, 5.

11. Noel Teachy and Warren G. Bennis. "Making the Tough Call," *INC. Magazine*, November 2007, 36-37.

12. Husted and West.

13. Ibid.

14. *Institute Report, 2007*.

15. Thomas A. Kolditz. *The Extremis Leader*, (San Francisco: Jossey-Bass, 2007), 6-18.

16. Ferrazza, Sal. Email to family, friends, and supporters, March 13, 2009.

17. James I. Robertson Jr. *Stonewall Jackson*, (New York: Macmillan, 1997), 191-192.

18. Stewart W. Husted. *George C. Marshall: Rubrics of Leadership*, (Carlisle: U.S. Army War College, 2006), 17.

19. Ibid, 20.

20. Ibid, 23-24.

21. Csorba. 24.

22. Len Marrella. *In Search of Ethics.* (Sanford: DC Press, 2001), 63.

23. Henry Wise. *Drawing Out the Man: The VMI Story,* (Charlottesville: University Press of Virginia, 1978), 64.

24. Francis H. Smith. *The Inner Life of the Virginia Military Institute Cadet,* (Lexington: VMI Board of Visitors, 1878).

25. Owen Connelly. *On War and Leadership,* (Princeton: Princeton Press, 2002), 122.

26. "Cheating is a Personal Foul," The Educational Testing Service/Ad Council Campaign to Discourage Academic Cheating, 1999.

27. Ibid.

28. Gina Cavallaro. "Cheaters Beware," *Army Times,* December 31, 2007, 8.

29. "Institutions of Liar Learning," *News & Daily Advance,* May 20, 2007, A7.

30. Reagan McMahon. "Everyone Does It," *San Francisco Chronicle,* September 9, 2007, 18.

31. Association of Graduates United States Air Force Academy, email, May 1, 2007.

32. Lawrence Kohlberg. *Cognitive-Development Approach to Moral Reasoning,* (Englewood Cliffs: Prentice Hall, 1981), 13-16.

33. Evan H. Offstein. *Stand Your Ground.* (Westport: Praeger, 2006), 32.

34. Charles C. Haynes and Marvin W. Berkowitz. "What Can Schools Do?" *USA Today,* February 20, 2007, 13A.

Chapter 6 Notes

1. Richard L. Hughes, Robert C. Ginnett, and Gordon J. Curphy. *Selections from Leadership: Enhancing the Lessons of Experience.* (New York: McGraw-Hill, 2001) 77.

2. "Cadet Michael N. Lokale '03 Named Rhodes Scholar," *VMI Alumni Review,* Winter 2003, 5.

3. "VMI Athletics." *VMI Alumni Review,* Summer 2006, 191.

4. Robert J. Sternberg. *Beyond IQ: A Triarchic Theory of Human Intelligence,* (New York: Cambridge University Press, 1985).

5. Ibid, 80.

6. Ibid.

7. Hughes, et al., 82.

8. "Defining and Assessing Learning: Exploring Competency-Based Initiatives," National Postsecondary Education Cooperative, September 2002.

9. L. J. Cronbach. *Essentials of Psychological Testing.* 4th ed. (San Francisco: Harper & Row, 1984).

10. F. E. Fiedler, and J. E. Garcia. *New Approaches to Leadership: Cognitive Resources and Organizational Performance,* (New York: Wiley, 1987.

11. *Vision 2039: Focus on Leadership,* 2006), 10.

12. Ibid.

13. *VMI Quality Enhancement Plan,* 2006.

14. Ibid.

15. Bob Holland. "Military Training in the Alps Tops Cadet's Study in France, *The Institute,* April 2008, 6.

16. "The Institute," *VMI Alumni Review,* Winter 2006, 147.

17. "Engineering Professor Named Virginia Faculty' Rising Star'," *VMI Alumni Review,* Spring 2005, 139.

18. "Environmental Virginia – VMI Research Labs," Undergraduate Research Initiative, www.vmi.edu/academics/show-apex, 2009.

19. Ibid.

20. *VMI Alumni Review.* 139.

21. Scott Belliveau. "A Conversation with Robert L. McDowell '68," *VMI Alumni Review,* Spring 2005, 136-138.

Chapter 7 Notes

1. Devin Millson. "Ease Up Brass, We're Exhausted," *The Cadet,* February 1, 2008, 3.

2. Hayes Johnson and George C. Wilson. *Army in Anguish.* (Washington: Washington Post, 1972), 92.

3. Geoffrey Norman. *The Institute.* (Wellington: Edgeworth Editions, 1997), 100-112.

4. "The Institute." *VMI Alumni Review,* Summer 2006, 188.

5. Wendy Lovell. "VMI Named VMI's First Goldwater Scholar," *The Institute Report*, April 2007, 5.
6. Bob Holland. "Deployed Cadets Welcomed Back to VMI," *The Institute Report*, October 2007, 5.
7. Jay Conley. "VMI Cadet Puts Studies to Test," *Roanoke Times*, July 25, 2007.
8. Henry A. Wise. *Drawing Out the Man: The VMI Story*, (Charlottesville: UVA Press, 1978) 104-105.
9. Paul N. Kotakis and Robert E. Wagner. "Introducing Cadet Command," May 1, 2008. Personal correspondence and notebook.
10. Dean Danas. "AF ROTC Cadets in Hot Pursuit of Flying Slots," *Institute Report*, February 2005, 15.

Chapter 8 Notes

1. Howard G. Hass, *The Leader Within*, (New York: HarperBusiness, 1992), 11-12.
2. Ibid, 53.
3. Warren G. Bennis and Robert J. Thomas. *Geeks and Geezers*, (Cambridge: Harvard Business School Press, 2002), 14.
4. *American Heritage Dictionary*, 2002.
5. Ibid, 74.
6. Laura F. Brodie. *Breaking Out*, (New York: Pantheon Books, 2000), 226.
7. Henry A. Wise. *Drawing Out the Man: The VMI Story*, (Charlottesville: UVA Press, 1978).
8. Ibid., 120.
9. Matt Chittum. "Va. Anti-Hazing Law 'Isn't Worth the Paper It's Written On," *Roanoke Times*, March 9, 1998.
10. "In Campus Clubs, Hazing Appears Common," *USA Today*, March 12, 2008, 7D.
11. Corey Hajim. "The Top Companies for Leaders," *Fortune*, October 1, 2007, 122-123.
12. Gregory Moorhead and Ricky W. Griffin. *Organizational Behavior*, (Boston: Houghton Mifflin Company, 2001), 558.

13. Robert B. Flowers. "On Being an Engineer: Reflections from the Chief of Engineers," *VMI Alumni Review*, Spring 2004, 9-10.

14. Doug Crandall. *Leadership Lessons from West Point*, (San Francisco: Jossey-Bass, 2007), 319.

15. Ibid, 321.

16. Wise, 287.

17. Ibid, 289.

18. Ronald D. Abbit. "College Orientation Workshop (COW) at VMI," *VMI Alumni Review*, Spring 2006, 142-142.

19. Ibid, 144.

20. "McDew '82, USAF, Promoted to Brigadier General," *VMI Alumni Review*, Summer 2006, 179-180.

21. "Anthony Q. McIntosh '89," *VMI Alumni Review*, Fall 2006, 165.

22. "Alumni News," *VMI Alumni Review*, Fall 2005, 144-145.

23. Alumni News, *VMI Alumni Review*, Issue 1, 2008, 184-185.

24. Brodie, 11.

25. Bob Brown. "Women at VMI," *Richmond-Times Dispatch*, July 22, 2007.

Chapter 9 Notes

1. James I. Robertson. *Stonewall Jackson's Book of Maxims*, (Nashville: Cumberland House, 2002), 86.

2. Bradford A. Wineman. "Prepared for the Varied Work of Civil Life," *VMI Alumni Review*, Spring 2003, 22.

3. Bruce E. Barber, *No Excuse Leadership: Lessons from the U.S. Army's Elite Rangers*. (Hoboken: John Wiley & Sons, 2000), 21.

4. John P. Jumper. "Gen. John Jumper '66, Chief of Staff, United States Air Force, Speaks at 2001 Institute Society Dinner," *VMI Alumni Review*, Winter 2002, 7-8.

5. Betsy Hiel. "Raider Brigade Hot on Saddam's Trail," *The Pittsburgh Tribune-Review*, December 15, 2003.

6. William A. Cohen. *The New Art of the Leader*. (Paramus: Prentice-Hall Books, 2000), 145.

7. Isaac Moore. "A Letter from Karbala," *VMI Alumni Review*, Fall 2003, 13.

8. Health Resources. "Fitness," www.healthresources.com, October 5, 2007.
9. W. Edwards Deming Quoted in *The Deming Management Method* by Mary Walton. (New York: Perigee Books, 1986), 73.
10. Larry R. Donnithorne, *The West Point Way of Leadership*, (New York: Currency Doubleday, 1994), 42-43.
11. Ibid.,44.
12. Harry E. Chambers and Robert E. Craft. *No Fear Management*, (Boca Raton: St.Lucie Press, 1998), 59-67.
13. Robertson, 18.
14. Glen Van Ekeren. *The Speaker's Sourcebook* . (Englewood Cliffs: Prentice-Hall, 1988), 148.
15. Doug Crandall. *Leadership Lessons from West Point*, (San Francisio: Jossey-Bass, 2007), 18.
16. Thomas Davis. *The Corps Roots the Loudest: The History of VMI Athletics*, (Charlottesville: UVA Press, 1986), 9.
17. Ibid.,66.
18. "Bobby Ross '59 Named Head Football Coach at Army," *VMI Alumni Review*, Winter 2004, 154-155.
19. Davis, 21-23.
20. Ibid, 20.
21. E-mail message from Donny White, January 2008.
22. Bert Williams. "Why Coach Reid Left," *The Cadet*, February 1, 2008, 2.
23. "Meet Jim Reid, Head football coach," *Keydet News*, Spring 2006, 2.
24. "Meet Jim Reed," 2.
25. Davis, 141-142.
26. Ibid, 21-23.
27. Ibid., 149.
28. *VMI 2003 Basketball Media Guide*, 8-9.
29. VMI basketball web page, 2008-2009.
30. Ibid.
31. Ibid.
32. "What They Are saying about VMI's Win at Kentucky," Keydet Club web page, www.vmi.edu/keydetclub, February 2009.

Chapter 10 Notes

1. Thomas Stonewall Jackson.
2. Glen Van Ekeren. *The Speaker's Sourcebook*. (Englewood Cliffs: Prentice-Hall, 1988), 242.
3. J. Thomas Wren and Marc J. Swatez. *The Leaders Companion*. (New York: The Free Press, 1995), 247.
4. Ibid., 248.
5. Richard E. Neustadt and Ernest R. May. *Thinking in Time*. (New York: The Free Press, 1986), 251-252.
6. John F. Kennedy. *Profiles in Courage*. (New York: Perennial, 1956).
7. David Freedman. *Corps Business*. (New York: Harper Business, 2000), 8-9.
8. John C. Maxwell. *Developing the Leader Within You*. (Nashville:Thomas Nelson, 1993), 83.
9. William A. Cohen. *The New Art of the Leader*. (Paramus: Prentice Hall Press, 2000).

BIBLIOGRAPHY

Axelrod, Alan. *Patton on Leadership*. Paramus: Prentice Hall, 1999.

Barber, Bruce E. *No Excuse Leadership: Lessons from the US. Army's Elite Rangers*. Hoboken: John Wiley & Sons, 2000.

Bennis Warren. "The Leadership Advantage," *Leader to Leader*. Leader to Leader Institute, Spring, 2003.

Bennis, Warren G. and Thomas Robert J. *Geeks and Geezers*. Cambridge: Harvard Business School Press, 2002.

Brodie, Laura F. *Breaking Out*. New York: Pantheon Books, 2000.

Byrd, Richard E. *Alone: The Greatest Antarctic Adventure of Our Time*. Island Press (paperback), 2003.

Chambers, Harry E. and Craft, Robert E., No *Fear Management*. Boca Raton: St. Lucie Press, 1998.

Cohen, William A. *The New Art of the Leader*. Paramus: Prentice-Hall, 2000.

Connelly, Owen. *On War and Leadership*. Princeton: Princeton Press, 2002.

Cronbach, L.J. *Essentials of Psychological Testing*. 4th ed. San Francisco: Harper & Row, 1984.

Csorba Les T. *Trust: The One Thing That Makes or Breaks a Leader*. Nashville: Thomas Nelson, 2000.

Cranwell, Doug. *Leadership Lessons from West Point*, New York: John Wiley & Sons, 2007.

Ekeren, Glen Van. *The Speaker's Sourcebook*. Englewood Cliffs: Prentice-Hall, 1988.

Davis, Thomas. *The Corps Roots the Loudest: The History of VMI Athletics*. Charlottesville: UV A Press, 1986.

Demming, W. Edwards. Quoted in *The Deming Management Method* by Mary Walton. New York: Perigee Books, 1986.

Donnathome, Larry R. *The West Point Way of Leadership*. New York: Currency/Doubleday, 1993.

Fiedler, F. E. and J. E. Garcia. *New Approaches to Leadership: Cognitive Resources and Organizational Performance*. New York: Wiley, 1987.

Freedman, David H. *Corps Business*. New York: HarperBusiness, 2000.

Gardner, Howard. *Frames of Mind: The Theory of Multiple Intelligences*. New York, Basic Books, 1983.

George, Bill and Sims, Peter. *True North: Discover Your Authentic Leadership*. San Francisco: Jossey-Bass, 2007.

Hass, Howard G. *The Leader Within*. New York: HarperBusiness, 1992.

Henslin, J. *Sociology: A Down-to-Earth Approach*. 7th ed., Needham Heights: Allyn & Bacon, 2005.

Hughes, Richard L., Ginnett, Robert C. and Curphy, Gordon J. *Selections from Leadership: Enhancing the Lessons of Experience*. New York: McGraw-Hill, 2001

Husted, Stewart W. *George C. Marshall: Rubrics of Leadership*. Carlisle: Army War College Foundation, 2006.

Husted, Stewart W. "The National D-Day Memorial: The Right Vision Becomes a Question of Fundraising Ethics," *Business Case Journal*, Winter 2007.

Kennedy, John F. *Profiles in Courage*. New York: Perennial, 1956.

Kohlberg, Lawrence. *Cognitive-Development Approach to Moral Reasoning*. Englewood Cliffs: Prentice Hall, 1981.

Kolditz, Thomas A. *The Extremis Leader*. San Francisco: Jossey-Bass, 2007.

Kouzes, James and Posner, Barry. *The Leadership Challenge: How to Get Extraordinary Things Done in Organizations*. 3rd. ed., San Francisco: Jossey-Bass, 2003.

Kouzes, James M. and Posner, James Z. *Credibility*. San Francisco: Jossey-Bass, 1993.

Manning, George and Curtis, Kent. *The Art of Leadership*. New York: McGraw-Hill, 2007.

Marrella, Len. *In Search of Ethics*. Sanford: DC Press, 2001.

Maxwell, John C. *Developing the Leader Within You*. Nashville: Thomas Nelson, 1993.

Moorhead, Gregory and Griffin, Ricky W. *Organizational Behavior*. Boston: Houghton Mifflin Company, 2001.

Neff, Thomas J. and Citrin, James M. *Lessons from the Top*. New York: Currency Doubleday, 1999.

Neustadt, Richard E. and May, Ernest R. *Thinking in Time*. New York: The Free Press, 1986.

Offstein, Evan H. *Stand Your Ground*. Westport: Praeger, 2006.

Robertson, James I. *Stonewall Jacktson's Book of Maxims*. Nashville: Cumberland House, 2002.

Smith, Francis H. *The Inner Life of the Virginia Military Institute Cadet*. Lexington: VMI Board of Visitors, 1878.

Stemberg, Robert J. *Beyond IQ: A Triarchic Theory of Human Intelligence*. New York: Cambridge University Press, 1985.

Teachy, Noel and Bennis Warren G. "Making the Tough Call," *INC. Magazine*. November 2007.

Wise, Henry. *Drawing Out the Man: The VMI Story*. Charlottesville: UV A Press, 1978.

Wren, Thomas. *The Leader's Companion: Insights on Leadership through the Ages*. New York: Free Press, 1995.

INDEX

ABOUT THE AUTHOR

Stewart Husted, a former business school dean, is the inaugural John and Jane Roberts Chair in Free Enterprise Business at the Virginia Military Institute. He earned his Ph.D. from Michigan State University, his M.Ed. from the University of Georgia, and his B.S. from Virginia Tech. An experienced speaker and writer (co-authored seven business texts including two best sellers and *George C. Marshall:*  *Rubrics of Leadership* (Army War College, 2006). Husted is also a retired Lt. Colonel in the U.S. Army Reserve and a Vietnam veteran. In addition, he is the founder of a non-profit organization (Goose Creek Adventure Learning, Inc.) devoted to educating and training youth and adults to become more effective leaders.

Printed in the United States
219428BV00004B/13/P

9 780982 017289